Electroplating and Related Processes

ELECTROPLATING
AND
RELATED PROCESSES

by

J. B. Mohler

Research Specialist
The Boeing Company

CHEMICAL PUBLISHING CO., INC.
New York 1969

Printed in the United States of America

INTRODUCTION

Electroplating was born from a science that demonstrated the remarkable ability of electric current to reduce metal salts to metal. It soon aided the production of beautiful objects and it became an art that was dependent on the masters who learned how to coax attractive coatings from homely solutions. But did the art of yesterday become the science of today? Or is the science of today the art of tomorrow? I think not. It will remain an art and a science. It is possible to design and operate an automatic plating process that is coldly technical. But it is not possible to remove personalities from the practice of electroplating.

Given a process, the individual will change it by science, logic, skill, intuition and art. Can anyone say that silver plating is best done with one strike, or two, or three. I prefer one. A friend prefers three. What is already known and communicated will help to understand but will not settle this difference. It is better to regard plating cycles as suggestions and plating baths as uncompleted formulations. The need for continuing experimentation is a vital part of every electroplating installation.

It is relatively easy to remove metal from solutions by the application of current, but only specific experimentally developed solutions produce useful electrodeposits.

If a metal salt is picked at random and current is applied to a solution of the salt, the results obtained will be varied. If the salt solution contains sodium chloride, only hydrogen will develop at the cathode. If a solution contains lead acetate, lead will deposit, but the deposit will appear as long crystals extending into the bath. If a solution contains stannic sulfate, the stannic ions will be reduced to stannous ions at the cathode and, at least for a short time and at a low current density, neither tin nor hydrogen will deposit. If a solution contains chromic acid, only hydrogen will develop; but if a small amount of sulfuric acid is added to the electrolyte, then chromium will deposit in addition to hydrogen.

By experimentation with solutions of metal salts, baths can be developed that will produce satisfactory deposits. A great many experiments are usually required and often it is necessary to study the effect of a host of organic substances known as addition agents. By experimentation and study, new baths have been developed to fill specific needs. For example, a deposit that is satisfactory for electrorefining may not be satisfactory for electroforming. A deposit that is satisfactory for electroforming may not be satisfactory for electroplating.

Electroplating is a process of electrodeposition by which a thin, smooth, sound metallic deposit is produced over a basis metal. This definition sets electroplating apart from the other processes of electrodeposition even though the requirements of the definition are not met in every case.

In this book, the fundamentals of electroplating will be briefly considered. A number of plating baths will then be discussed. It is hoped that the text will aid the practices of the science and the art of electroplating.

July 1968 *J. B. Mohler*

CONTENTS

Introduction v

1. Mechanism of Electrodeposition 1
2. Laws and Characteristics of Plating Baths 8
3. The Deposit 17
4. Preparatory Steps of Plating 25
5. Preparation of the Surface 31
6. Cleaning 43
7. Pickling 60
8. Strike Plating 68
9. Rinsing 70
10. Anodizing 78
11. Brass Plating 83
12. Bronze Plating 89
13. Cadmium Plating 95
14. Chromate Coatings 103
15. Chromium Plating 106
16. Acid Copper Plating 117
17. Copper Cyanide Baths 127
18. Iron Plating 136
19. Lead Plating 140
20. Lead–Tin 145
21. Nickel Plating 152
22. Electroless Nickel 163
23. Phosphate Coatings 167
24. Silver Plating 169
25. Acid Tin Plating 179
26. Alkaline Tin Plating 185
27. Tin–Nickel 192
28. Tin–Zinc 194
29. Acid Zinc Baths 198

30. Zinc Cyanide Baths 201
31. Control of a Plating Bath 208
32. Plating Tests 214
33. Gravity, Conductivity, and Voltage 219
34. Electroplated Alloys 229
35. Layer Plating 236
36. Applications of Electroplating 242
37. Plating Bath Troubles 253
38. Continuous Plating 258
39. Plating on Plastics 262
40. Preparation of Metals for Painting 266
41. Analytical Methods for Plating Baths 272

Appendix 293
Conversion Factors 293
Electrochemical Yields 294
Electrochemical Formulas 294
Electrochemical Equivalents 295
Single Electrode Potentials 295
Stripping Chart 296

Glossary 299
Index 305

1. MECHANISM OF ELECTRODEPOSITION

The Electrolyte

The process of electrodeposition is one in which electric current is carried across an electrolyte and in which a substance is deposited at one of the electrodes.

The electrolyte is the medium that carries the current by means of ions. The ability of a solvent, especially water, to ionize substances dissolved in it, i.e., to split them into components that carry positive and negative charges, makes electrolysis possible. The electricity is carried across the electrolyte by the charged ions and products of electrolysis appear at the electrodes. This is a result of the positively charged ions being attracted to the negatively charged cathode while the negatively charged ions travel toward the positively charged anode. The charges of the ions are then neutralized by the charges on the electrodes and the products of the electrolysis appear at the electrodes.

The electrolyte is a conducting medium in which the flow of electric current is accomplished by the movement of matter. It is also a substance that gives rise to ions. If more than one ion is present, carrying a positive charge, several reactions are possible at the negatively charged cathode, although usually only one product of electrolysis appears. Each electrode reaction takes place at a specific voltage and the most positive metal ion will deposit at the cathode.

Any liquid or solution that contains ions can be used as an electrolyte. The large majority of commercial electrolytes, however, use water as the solvent and are therefore called aqueous electrolytes. Fused salts, which are a class of nonaqueous electrolytes, find their greatest use in the electrolytic production of metals such as sodium, magnesium, and aluminum. Fused-salt electrolytes are also used in the electrolytic cleaning of metals.

The extensive use of water as the solvent in the electroplating industry is due to its cheapness and abundance and to the fact that many com-

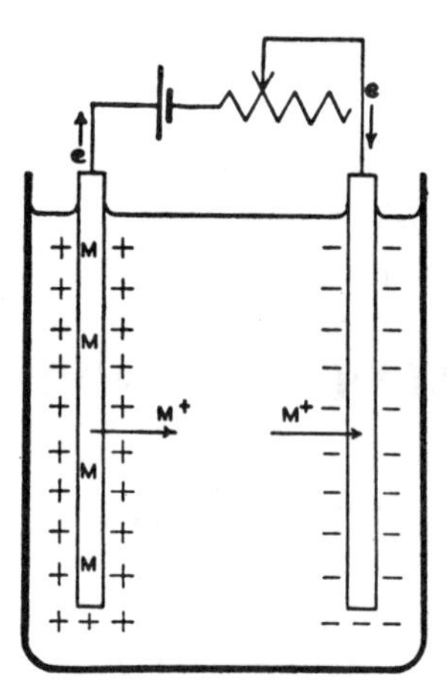

Fig. 1 Electrodeposition.

mon salts are very soluble in water.

In this book, the aqueous electrolytes alone will be considered. The term electroplating will occasionally be simplified to "plating."

The Cathode

The plater is primarily interested in the reaction that takes place at the cathode since this is where deposits are produced. The potential at which this reaction takes place is called the deposition potential. This potential can be measured readily in the laboratory, but it is neither convenient nor desirable to measure it in the plating tank. The reaction at the cathode is much easier to follow by a knowledge of the quantity of current that reaches the cathode. If, in addition, the plater has some knowledge as to the distribution of the current over the cathode, he may then have some idea of how the plated article will be coated with metal. He will be able to predict the time required to produce a desired thickness of deposit and also have an idea of how the thickness will vary from one area to another.

Unfortunately, it is difficult to make accurate predictions regarding the distribution of metal over the cathode. In practice, the quantity of current is controlled and the current is allowed to flow for a definite period of time, after which the local thickness of metal on the plated piece is measured. If an undesirable distribution of metal is obtained, adjustments are made in the racking or positioning of the pieces or of the anodes. In some cases, chemical adjustment of the electrolyte may improve the metal distribution.

The current allowed to flow to the cathode is in proportion to the area being plated, so that the current is expressed as current density or quantity of current per unit area. In commercial plating, the current density is expressed as amperes per square foot.

Occasionally, a plating tank is controlled by voltage. This procedure is less satisfactory than control by current density, since the tank voltage is affected by many factors other than the reaction taking place at the cathode. Tank voltage, however, is very easily measured and often gives information about changes in the plating process, such as a reduction in conducting salt content, or polarization of the anode. Plating control by tank voltage is satisfactory when the cathode area is difficult to measure, such as in barrel plating.

The Anode

The reactions taking place at the anode are almost independent of the reactions occurring at the cathode. The position of the anodes naturally has much to do with the distribution of current at the cathode, but the anodes usually operate best at a range of current densities that can be changed independently of the cathode current density by changing the anode area.

The Balanced Bath

A plating bath can be operated successfully for long periods of time if the composition of the bath is not changed too rapidly. Such changes are primarily due to:

1. Chemical decomposition.
2. Incomplete electrode reactions.
3. Drag-in or drag-out.

The bath stability can best be illustrated by consideration of several typical baths.

An acid copper bath is relatively easy to control because there is very little tendency for chemical decomposition and the reactions at the electrodes are essentially complete. This means that the electrolyte is chemically stable and that for every chemical equivalent of copper dissolved at the anode there is a chemical equivalent of copper deposited at the cathode. Nevertheless, the bath cannot be continuously operated without control, since solution is lost from the system by drag-out every time a rack is removed. Control is further complicated in that glue is generally added to the bath to produce a finely crystalline deposit. The glue is not stable and must be controlled. This bath is typical of many acid baths where the major factor in control is the addition agent used—in this case, the glue.

An alkaline tin bath is not a typical alkaline bath, but it is a good example of a bath where care is required to maintain solution balance.

In the alkaline tin bath, the electrode reactions are not complete, i.e. the anode and cathode efficiencies are less than 100%. In addition to depositing tin, hydrogen is evolved at the cathode. And in addition to tin being dissolved, oxygen is given off at the anode. To make the problem more complex, the bath undergoes continuous chemical decomposition and the anodes must be maintained with an oxide film at all times. If the oxide film is not present or is too thick, troubles set in that throw the bath out of balance or even cause a deposit to form that is not acceptable. The bath can be controlled by regulation of bath temperature, cathode-current density, anode-current density, and by chemical analysis and proper chemical additions. The bath voltage is responsive to changes at the anodes; and since the bath is sensitive to anode changes, the voltage may be used as an aid to bath control.

A plating bath should always be kept within prescribed chemical limits whether or not the bath composition is difficult to maintain. The bath should also be used in such a way that a minimum of chemical additions is required. It is very rare that a bath does not require frequent additions, although this condition is approached when the anode efficiency is slightly higher than the cathode efficiency. Such a bath would be a perfectly balanced bath, but even in this case, drag-out would eventually remove a sufficient amount of one of the essential chemicals so that chemical additions would be required. Since additions are required to all baths, it is best to make them frequently and in small amounts, so that chemical limits are easily held. Addition of large quantities of chemicals to the plating bath often leads to trouble. For example, with a large addition of chemicals, a small error in chemical analysis is liable to result in a concentration exceeding the chemical limits on the bath. Moreover, the chemicals often contain impurities that are not harmful for small additions but that require electrolysis with dummy cathodes before the bath may be used, if the additions are large.

The Equilibrium Potential

It is well to settle a few points regarding the equilibrium potential before considering some of the details of the plating process.

The well-known series of potentials for electrochemical reactions is shown in Table 1. This series has also been called the electrochemical series or the electromotive force series of elements. It is a reliable point for reference; however, it can be highly misleading if it is used as a definite guide. These potentials are equilibrium potentials. They were measured under conditions where no current was flowing and with definite quantities of dissolved salt present and at a standard temperature to obtain comparable valves.

Hydrogen is taken as the reference point on the electrochemical scale and is arbitrarily assigned a value of zero. The metal ions whose voltage is listed as positive are more reactive than the hydrogen ion when they are present in equivalent chemical quantities, whereas the metal ions having a negative voltage are less reactive than the hydrogen ion. Thus, as we go up the scale, the metal ions become more and more reactive at the cathode, i.e. they become more electropositive (attracted to the cathode), or they deposit more readily. As we go down the scale to more negative electrode potentials, the metals become more electronegative, or they go into solution more readily

Table 1 Electrochemical Potential Series

Ion	*Voltage*
Au^{+}	+1.5
Ag^{+}	+0.7995
Cu^{+}	+0.0528
Sn^{++++}	+0.003
H^{+}	0
Pb^{++}	−0.1264
Sn^{++}	−0.1406
Ni^{++}	−0.231
In^{+}	−0.336
Cd^{++}	−0.4024
Cr^{+++}	−0.509
Zn^{++}	−0.762
Na^{+}	−2.7125

(anodic metals).

With reference to this scale, it has been said that any metal on the scale will displace from solution all those metals that appear above it. This is true for the conditions under which the potentials were measured, but it cannot be taken as a general rule. According to this rule, if a piece of copper is immersed in a solution containing silver ions, silver will deposit on the copper. This will occur in acid solutions even though the solution contains a very small quantity of silver. However, the opposite reaction can be made to take place. That is, if silver is immersed in a concentrated acidic solution of copper, copper will deposit on the silver. In other acid solutions tin can be deposited on copper[1] even though tin is below copper on the electrochemical scale. These apparent exceptions occur because the conditions under which the experiments are carried out differ greatly from those under which the potentials were measured. However, the metals in the series are arranged according to their relative reactivity. The metals below hydrogen can be dissolved readily in acids and those very low in the table will react with water. The metals above hydrogen can only be dissolved in acids under oxidizing conditions. Thus, the table indicates chemical reactivity, but if it is to be applied to new or unusual conditions, it will be necessary to experiment or measure the potential under the new conditions. When the potential is measured during plating, it is measured under dynamic conditions and is called a deposition potential.

The Deposition Potential

The potential of an electrode and of a solution of its ions may be measured during plating. This deposition potential varies with the concentration of metal ions in the bath and is also greatly affected by the current density. As the current density is increased, polarization at the electrode increases, resulting in conditions more favorable for deposition of metals low in the electrochemical scale. Thus, it becomes possible to deposit zinc from acid solutions, whereas zinc normally dissolves in acids. One might expect that zinc could not be deposited in the presence of hydrogen ions, and under some conditions it is difficult to achieve this. If polarization does not exceed the hydrogen overvoltage, hydrogen will develop exclusive of zinc. This can take place even during the electrolysis of copper from a copper nitrate solution. Such an electrolysis is carried out in the presence of nitric acid, and by continued electrolysis, the copper can be deposited completely from the solution. However, if the solution is heated while copper is depositing, a point will be reached where the tendency for the copper to dissolve will be greater than its tendency to plate. At this point the copper will go back into solution in the nitric acid even though the current is flowing. The behavior of copper in nitric acid solutions and that of zinc in acid solutions illustrate that to study the mechanism of deposition the measurement of deposition potentials is helpful. These potentials direct attention to the importance of polarization, overvoltage, chemical reactivity, and ability to plate in the presence of other ions.

REFERENCE

1. U.S. Patent 2,369,620.

2. LAWS AND CHARACTERISTICS OF PLATING BATHS

The plater is concerned with the quantity of electricity that is used in his process, since the number of units of electricity that pass through the electrolyte during electrolysis determines the quantity of product formed at the cathode.

The unit quantity of electricity is the coulomb, which is the quantity of current passing in a circuit when 1 ampere flows for 1 second. Ordinarily, the ampere rather than the coulomb is considered in electroplating. The ampere, however, gives only the current strength and requires a time factor, so that the quantity of metal obtained at the cathode is calculated from the ampere-seconds or ampere-hours.

One gram equivalent weight of ions will be discharged at the cathode by 1 faraday. One faraday equals 96,500 coulombs, 96,500 ampere-seconds, 1608.4 ampere-minutes, or 26.806 ampere-hours.

One gram equivalent weight of a metal equals 1 gram atomic weight divided by the valence. The valence is equal to the number of charges on the ion. From the faraday, the grams deposited at the cathode can be calculated as in Table 2.

The quantity of electricity that passes through the electrolyte during electrolysis is important because a minimum amount of metal is required for every application. Also it may be desired to avoid an ex-

Table 2 Calculation of the Grams of Metal Deposited at the Cathode

Metal	*Ion*	*Gram atomic weight*	*Valence*	*Grams deposited per faraday*
Silver	Ag^{+}	107.88	1	107.88
Copper	Cu^{+}	63.57	1	63.57
Copper	Cu^{++}	63.57	2	31.78
Gold	Au^{+}	197.20	1	197.20
Tin	Sn^{++++}	118.70	4	29.67

cess of metal for economic or specification purposes. However, it is not enough to know the total quantity of current that passes. The distribution of the current and the plating efficiency must also be known for a reasonable estimate of the thickness. If the piece to be plated is a large flat sheet, then an excess of 10 to 20% of the current may be lost to the edges. The current density over most of the area will then be 10-20% lower than the current density obtained by dividing the total current by the total area. From this approximate current density, the approximate plating rate can be calculated. In many cases, however, the cathode efficiency is not 100% because a part of the total current is consumed in evolving hydrogen.

The cathode efficiency is that percentage of the current that deposits metal. In other words, if 60% of the current deposits metal, then the plating rate is only 60% of that which would be obtained by calculation from the ampere-hours. Just as it is more convenient to use current density than total current, it is also convenient to use ampere-hours per square foot to calculate plating thickness. Of course, if the cathode efficiency is less than 100%, then the plating rate is calculated from the following equation:

$$\text{plating rate} = \frac{\text{amp-hr}}{\text{sq ft}} \times \%\ \text{efficiency} \qquad (1)$$

The plating rate calculated from equation (1) is in ampere-hours per square foot. In order to convert this to thickness, the chemical equivalent and the specific gravity of the metal must be taken into consideration. This can be done with the following formulas:

$$\frac{\text{equivalent weight}}{\text{amp-hr faraday}} = \frac{\text{equivalent weight}}{26.8}$$

$$= \frac{\text{g}}{\text{amp-hr}} \text{ at } 100\%\ \text{efficiency} \qquad (2)$$

$$\frac{\text{g}}{\text{amp-hr}} \times \frac{\text{amp-hr*}}{\text{sq ft}} = \frac{\text{g}}{\text{sq ft}}$$

$$\frac{\text{g}}{\text{sq ft} \times 144} \times \frac{1{,}000}{16.4 \times \text{sp.gr.}} = \text{mils thickness} \qquad (3)$$

These formulas illustrate how the total current produces a total

* Plating rate from equation (1).

thickness. In every application, it is convenient to know the ampere-hours per square foot and the cathode efficiency. Not only is it convenient to have this information for the average current density, but it is also desirable to have it for the areas of highest and lowest current density. If the distribution of current is known and if a specification for thickness is given, then the required quantity of current can be determined.

Unfortunately, the distribution of a deposit over the surface of an article is not easily estimated and in many processes, it is impossible to calculate. Nor can the current density always be calculated. This is especially true of barrel-plating as well as operations in which the amount of current being taken by the plating rack is unknown. When the plating rate then *cannot* be calculated it must be determined by measurement of the plate thickness, after plating at a controlled period of time and a total current per rack or per barrel.

Experimentation in the plating room to obtain proper metal distribution should be guided by a knowledge of the qualitative rules for the behavior of the cathode.

These qualitative rules may be stated as follows:

1. The current concentrates on corners and edges.
2. The current does not penetrate into deep recesses.
3. The current does not readily flow around a nonconductor.
4. One conductor will rob another of current.

These rules may be used as a guide to the arrangement or geometry of a plating system.

Electrode Position

Rule 1 tells us that excessive current may be lost to projections or edges of the article. We also know that current will tend to follow the shortest path between conductors. This fact can be turned to advantage in order to reduce current concentration on edges by placing the edges farthest from the anode. Thus, in plating from an anode to a sheet, the system of Figure 2 can be rearranged to that of Figure 3.

According to Rule 2, if it is desired to plate inside of a large recess, the obvious thing to do is to place an anode in the recess. The anode should be shaped so that the anode-to-cathode distance is the same at all points over the surface and a little longer to the edges.

Fig. 2 Anode arrangement to favor plating on corners.

Fig. 3 Anode arrangement to minimize plating on corners.

Shadowing

A problem cannot always be solved by the proper arrangement of the system. Even when a solution is possible, it is not always economical. Anodes are difficult to support in fixed position with respect to the cathode and they change in size as they are used.

The edges of a sheet may be shadowed by the use of a nonconductor. The shadow does not have to be between the anode and the cathode but merely protect the cathode from receiving current in more than one direction. If a flat sheet cathode is placed between two nonconducting walls, the distribution of current over the cathode will be almost perfect, as shown in Figure 4. The walls prevent the current from coming from any other direction than in straight lines within the boundaries of the insulating walls.

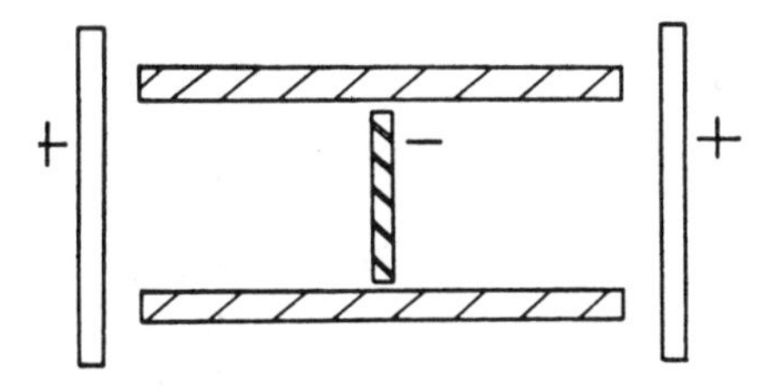

Fig. 4 Shadowing.

Robbing

If several flat sheets are plated at the same time, one may be used to

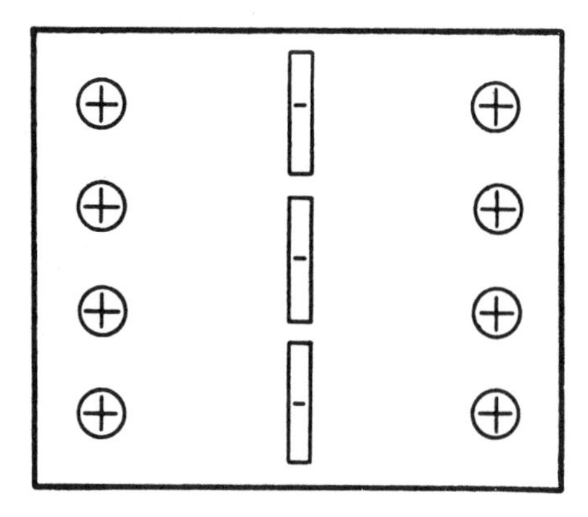

Fig. 5 Robbing.

rob the edges of the next. Fair uniformity may be obtained by the arrangement shown in Figure 5.

Here the sheets are close together and the edge of one robs current from the next. The edges next to the tank are shadowed by the walls of the tank provided the tank is a nonconductor. If the tank is a conductor, then the shortest electrical path may be from the anode to the tank and from the tank to the cathode. This condition often occurs in steel tanks and is corrected either by increasing the tank-to-electrode distance or by hanging a nonconductor, such as a sheet of rubber or a sheet of glass, between the tank and the electrode.

The qualitative rules have been illustrated with flat sheets. The same generalizations apply to pieces of complicated design. High-current-density areas should be robbed or shadowed. Changes should be made to get more current to low-current-density areas either by making the anode-to-cathode distance more favorable or by shadowing the high-current-density areas.

Polarization

"Polarization" is a term that is familiar to the electroplater. There are many electrode reactions that can be explained by means of it.

Polarization is the result of a dynamic condition at the cathode, due to changes in concentration brought about by the movement and discharge of ions. Because of the dynamic condition, a force is set up that tends to resist the flow of current. This force can be measured as voltage and is called polarization. Polarization is always present during electrolysis. Normally, the plater does not concern himself too much with it except when it becomes excessive and troubles are encountered due to the formation of undesirable products at the cathode. An exceptional case is that of tin anodes in a sodium stannate bath. Here the anodes must be maintained in a state of excessive polarization or difficulties will arise both in the bath and the plate. In this case, polarization is excessive. It is more common to refer to polarization when one really means "excessive polarization."

The electrode reactions can best be explained by the use of an ideal current-voltage curve, as shown in Figure 6.

If the voltage from the cathode to a point a very short distance from the cathode is measured under ideal conditions, a curve as shown in Figure 6 is obtained. At point E, no current is flowing so that the voltage is that of the equilibrium potential. Now, if a small amount of current is applied to the system, the voltage will increase and polarization will begin to take place. However, the deposition of metal will not yet occur even though there are platable ions in the aqueous solution. As the current is further increased, the voltage also increases until a point is reached at which plating takes place. This voltage is point D, known as the decomposition potential or the point at which the electrode reactions begins. If the solution consists of water and sulfuric acid, bubbles of hydrogen gas will appear at the cathode at

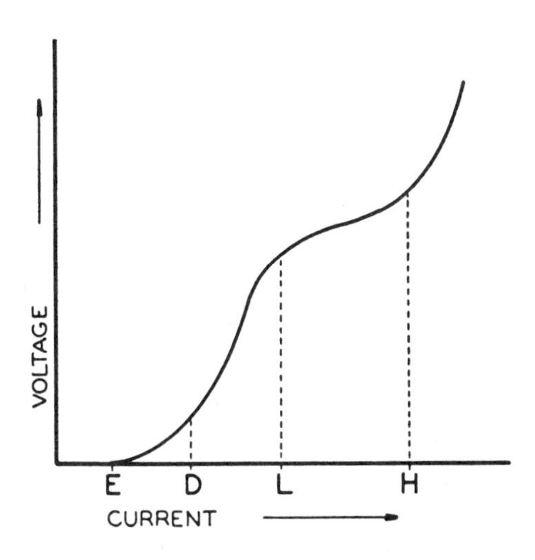

Fig. 6 Current-voltage curve.

this point. The voltage in this case is called the hydrogen overvoltage. (Hydrogen overvoltage is merely a form of polarization.)

If the solution contains sulfuric acid and copper sulfate in water, a decomposition potential will be reached at which copper will begin to deposit exclusively. As long as copper ions are supplied to the cathode at a rate providing an excess, this exclusive plating of copper will continue. With increase in current, the voltage will increase but not to a great extent. But with a continued increase in current, a point will be reached at which the copper ions are being discharged as fast as they are brought up to the cathode by diffusion. This will be point L on the ideal curve. The current density at this point is known as the limiting current density. For current densities greater than this, the cathode efficiency will drop and, in many cases, the plate will become rough or spongy even though the amount of hydrogen evolved is small. If the current is increased beyond the point of limiting current density, the voltage will increase more rapidly with increase in current until the point H is reached. At this point hydrogen will be evolved rapidly and the hydrogen overvoltage will have been exceeded with the copper ion still present but not in sufficient concentration to deposit exclusively. Beyond point H, the voltage will not increase greatly with increase of current. This is due to the aqueous solution having a very great capacity to supply hydrogen ions as fast as they are discharged.

Plating Quality

The quality of plating varies with the application. If the only purpose of the deposit is to prevent rusting of steel for a short period of time, then a very thin coating of zinc will be sufficient. The deposit will be a quality deposit if it completely covers the piece. If it is further required that the deposit be bright, then the job of quality plating becomes a little more exacting.

The quality of thin deposits may be controlled by testing thickness and corrosion resistance. As long as representative samples pass these tests, the specifications are met. The specifications, however, become more rigid with more exacting applications. As an example, a heavy silver deposit for an aircraft bearing must be almost flawless. The deposit must be bonded to the steel back and must be practically free from pores down to a microscopic size. The finished bearing must

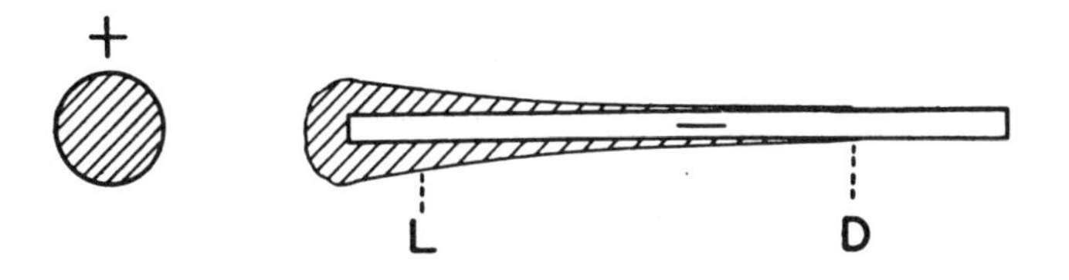

Fig. 7 Plating range.

meet strict porosity and adherence requirements. Here, quality has an entirely different meaning. And to produce the desired quality, very close control is required throughout the entire plating cycle.

In an exacting engineering application, quality may only be obtained by operating a plating bath well above the decomposition potential and well below the limiting current density. It may further be required that the bath be continuously agitated and heated, and that temperature, chemical limits, and the geometry of the plating system be closely controlled. Wide limits may be possible in some of the bath variables, but only if one variable is changed at a time. Therefore, close limits are set on all variables so that if all change in an unfavorable direction at the same time, the bath will still produce quality plating. In order to set such limits, some knowledge of the plating range is required.

Plating Range

The plating range is the range of current density over which the deposit will be satisfactory. It is usually a range of current densities somewhat above that at the decomposition potential and somewhat below the limiting current density. It will be very wide in an alkaline tin bath and quite narrow in a chromic acid bath.

In Figure 7, plating takes place from a bar anode to a sheet in which the edge of the sheet faces the anode. A section of the system is shown for a deposit from an acid bath of fair throwing power. On the edge of the sheet facing the anode the deposit is rough. The current density on this edge is very high and becomes less at greater distances from the anode. Point L is the limiting current density and point D the current density at the decomposition potential. The deposit is often rough, crystalline, off color or spotty at current densities that are very high or very low. Below some maximum and above some minimum current densities, the deposit is satisfactory and the range between the densities is the plating range.

The plating range can be measured experimentally by calibrating an arrangement of the anode and cathode. The two most common of these tests are the bent-cathode test[1] and the Hull cell test.[2]

REFERENCES

1. Pinner and Baker, *Trans. Electrochem. Soc.*, 77, 353 (1929).
2. F. MacIntyre and R. O. Hull, *Convention Proc. Am. Electroplaters Soc.* (1943).

3. THE DEPOSIT

An experienced plater can control a bath for long periods of time merely by observing the appearance of the deposit. He is quick to recognize the need for additions by appearance of dull areas on the plated work. He knows that red copper deposits mean low free cyanide in a copper bath. Furthermore, he knows the proper addition and current changes to bring about an immediate correction. Although such practice is good in emergencies, it is not recommended. But it does demonstrate that the product is the deposit and that it is possible, if not desirable, to continue to produce a good product by observation and remedy.

The Environment

The deposit is mostly affected by the environment in the immediate vicinity of the cathode. It is not affected by dirt that lies on the bottom of the tank or by sludge that adheres to the anode. It is influenced by tank walls that are close to the work and by suspended charged particles if they are close by. With time, the plate may occlude dirt or sludge that becomes suspended in the bath and reaches the deposit, but the condition at the cathode is somewhat independent of the composition of the bath and of the anodes. Identical deposits can be produced in baths that are entirely different by adjustment of current, temperature or agitation. Identical current distribution can be attained by changing the position of the anodes and using shadows to compensate for the change in anode position.

Normally, a bath is run in such a fashion that changes in the solution and in the anodes are immediately reflected by a change in the deposit. In this respect, it can be said that under normal operation, the cathode is dependent on solution changes and on proper anode control. However, it is good to know that the cathode may be treated almost independently of the bath and the anodes.

To illustrate the preceding points, let us assume that a Rochelle copper bath is operated at 140°F. Further, in order to solve a racking problem, a wax is used that softens above 125°F. Now, by decreasing the free cyanide and increasing the current density, it is possible to maintain the same quality at 120°F. New limits are set on the bath and the same result is obtained. The new limits may not be as desirable as the old ones because if the bath is operated at lower temperatures and a lower free cyanide content, the anode solubility may not be as satisfactory. Nevertheless, the bath will operate perfectly with more frequent chemical adjustment and when the wax racking method is dispensed with, the bath can be returned to the old limits. Almost any bath may be temporarily changed in a similar fashion.

When a nickel bath is being used for plating on steel, it is often desirable to operate the bath at low pH since the pH is more easily controlled in the low range. If the nickel bath is to be used temporarily to plate zinc-base die castings, it will be desirable to operate at high pH in order to reduce the rate of attack of the zinc alloy and thus reduce the rate of contamination. Nickel carbonate may be added to the bath to increase the pH, and by changes in the boric acid content and current density, a satisfactory plate can be obtained.

Baths should not be changed without experimentation to determine what compensating changes are required in chemicals, current density, temperature, or agitation. The greatest danger is that the plating range will be reduced. But if plating-range tests are made and the changes in the bath are not radical, changes may be made to meet special plating requirements. Experimentation should be preceded by a study of the published chemistry of the bath to appreciate the function of each bath ingredient and what compensating changes might be made. Also, it will be found that some changes are impossible. If it is desired to operate an alkaline tin bath at low temperature, it will be found that the cathode efficiency drops rapidly as the temperature is reduced and that there are no adjustments that can be made to make the bath plate tin at room temperature.

Properties of the Deposit

The science of electrodeposition is a branch of electrochemistry and a knowledge of chemistry is really essential to bath formulation.

However, the product is a metal that belongs to the realm of the metallurgist.

The conditions under which metals are produced by electrodeposition are very different from casting methods, but the deposit is, of course, governed by the laws for the metals. The fundamental difference between plating and casting is that in the first process the metal is formed very slowly and is formed cold as compared to the casting process. In general, the grain size is very much smaller than that in cast metals and it is to be expected that the deposit will have the properties associated with small grain size. The grain size is often so small that it cannot be resolved under the microscope. Because of this, electrodeposits are similar to cast metals after extreme cold working. The deposits will, in many cases, be hard and brittle due to the fine grain size, but, of course, these properties can be changed by heat treatment. Also, by changing the bath composition, deposits can be formed that are soft and ductile.

For each application, certain physical properties of the deposit are required. Before measurements are made, it is well to consider the mechanism of formation of the deposits as their properties will be governed by this mechanism.

The standard methods of measurement are not readily adaptable to the evaluation of the deposit. Methods should be used that can be applied to thin sheets as the deposit is usually quite thin and the properties of the thin deposit may be different from those of a heavy section that was produced merely for testing. The first layers of metal that are laid down are often finer-grained than later layers so that the tensile strength may decrease as the deposit becomes thicker. In order to obtain a specimen, a parting compound, such as graphite or a very thin layer of grease, may be applied to the cathode. Or, if the cathode is buffed and cleaned with solvent or magnesia but not etched, it is often possible to strip the deposit from the cathode. However, a deposit that is formed over a buffed cathode is liable to have a finer grain than one that is formed over a crystalline or etched substrate. Metallographic examination will reveal how the grain size of deposited metal is influenced by the basis metal. If acid copper is deposited on clean etched copper, it will tend to reproduce the grain size of the underlying cast copper. If, however, acid copper is deposited on buffed copper, or any buffed metal, even though the metal is etched, it will tend to reproduce the fine grain of the cold-worked surface. As the plate grows, the grain size will change until a thick deposit will produce grains that are characteristic of the bath regardless of the influence of the first layers of metal.

If the electroplating process is regarded as cold casting, it may then be expected that many properties of the deposit will not correspond to those of a metal produced by the hot methods of a foundry.

Crystal Structure

The crystal structure of cast metals is best observed by etching. Examination of the cast surface is not very revealing because surface effects due to oxidation and freezing of the surface obscure the manner in which the crystals grew. An electrodeposit on the other hand can be studied at any point in the process by merely removing the cathode from the bath and examining it. Examination often reveals the crystalline nature of the surface. If the deposit is visibly crystalline, it is known as a "crystalline" deposit. If the deposit has to be examined under the microscope to detect its crystalline nature, then it is "finely crystalline." Surface examination under the microscope may reveal that the deposit is crystalline, nodular, or quite flat. Often, if the deposits are heavy, they will present a macroscopic appearance similar to the appearance of thinner deposits under the microscope.

Crystalline deposits and even large crystals of metal growing out into the bath are characteristic of simple pure metal salts. Pure silver nitrate, copper sulfate, or lead acetate can produce large loose crystals that are of no practical value. However, secondary salts or foreign substances introduced into the bath may break up the simple crystalline structure. Sulfuric acid will have such an effect when it is added to copper sulfate. By addition of glue, the deposit may become even more finely crystalline and sufficiently flat to be bright. All impurities do not produce this effect. Some promote growth perpendicularly to the basis metal and long needles or leaved deposits are formed. By experimentation, additives have been developed for many of the baths, which produce a finely crystalline, a smooth, or a bright deposit. The addition agents include a host of organic compounds, colloidal substances, wetting agents, and heavy metal salts. An approximate definition of addition agents is as follows: Addition agents are substances that improve the quality of the deposit when added in small amounts. Such a definition would include salts, organics and noble metals. The primary salt, is the salt supplying the ions of the metal being plated.

In alkaline baths, basic salts of the metal trapped in the deposit produce an addition-agent effect. In general, the alkaline baths are more prone to produce sound deposits without the aid of addition agents, although additions are used to obtain bright deposits.

In view of the fact that the deposit will trap metals, metal compounds, and organic substances, it may consist of a pure metal, solid solutions, intermetallic compounds, two or more metallic phases or of metal and oxides, sulfides or organic substances. Usually the amount of oxygen, sulfur or carbon present, as the elements of an organic or inorganic compound, is very small—of the order of a few hundredths or a few thousandths of a percent. Therefore, the analysis of the deposit will usually be 99+% pure metal. If a second metal is present, it may run as high as several percent before it will cause any noticeable difference in the appearance of the deposit.

The crystal form of the deposit is usually in the crystal system that is normal for the metal. Occasionally, the basis metal influences the deposit to such an extent that a crystal form is produced that is abnormal for the particular metal. Electrodeposited cobalt[1] and chromium[2] have been shown to be deposited in forms not produced by metallurgical means.

Recrystallization

Because of the small grain size of electrodeposited metals, recrystallization usually yields small equiaxed grains, but the recrystallized grain size is influenced by the original grain size of the deposit. Occasionally, unusual effects may be produced due to the presence of small amounts of impurities that migrate to grain boundaries on recrystallization.

Stress

Metals can be deposited under extreme stress. Forces can be trapped in the metal that are so severe that they will tear the deposit from the basis metal. Although extreme stress is unusual, very high stress is normal for deposits of chromium or hard nickel. For example, if chromium is deposited on one side of a thin metallic sheet, the sheet will bend. If the chromium is then etched under proper conditions, the stress will be relieved and the sheet will straighten. The amount of stress can be varied by change in bath composition and by change in plating conditions. In popular bath formulations and normal applications, stress causes little trouble, but with thin sheets difficulties arise. If a highly stressed deposit is applied to both sides of a thin sheet, the sheet will not bend because the force will be equal on both sides. But the stress can be great enough to break thin sheets, and this effect, known as the "sandwich effect," is sometimes confused with hydrogen embrittlement.

Hydrogen Embrittlement

Atomic hydrogen can cause embrittlement of the deposit or the basis metal. The hydrogen given off at the cathode is in such atomic or "active" form. This takes place in electrolytic cathodic alkali cleaning, in pickling, and in plating. The evolved hydrogen can be eliminated in the cleaning and pickling steps by using anodic treatment and the embrittlement in the plating step can be reduced by changing the plating conditions. The embrittlement may also be relieved by heating after plating. This is common practice in hard chromium plating.

Hard chromium is deposited to produce long-wearing surfaces, but the deposit is often so brittle that it will crack on grinding. The brittleness can be relieved by heat treatment after the plating operation. Although hydrogen may have considerable effect on the ductility, it has little effect on the hardness of chromium.

When high strength steels are electroplated they become embrittled and may fail in service. This becomes a factor of concern with steels of 220,000 psi ultimate strength and higher. Steels that have been chromium plated can be relieved of embrittlement by heating to 375°F for 3 or more hours. Chromium is permeable to the embrittling hydrogen, particularly the common chromium plate that is permeated with cracks. Cadmium, on the other hand, traps hydrogen so that heating is not successful when excessive hydrogen has been introduced into the steel before or during cadmium plating. In order to produce non-embrittled cadmium plated high strength steels it is necessary to avoid cleaning processes that introduce hydrogen, to restrict pickling to a short, non-gassing acid dip, and to employ special low embrittling cadmium plating processes. Finally the plated part should be baked up to 23 hours at 375°F.

Table 3 The Hardness of the Deposit

Metal	Type of solution	*Electrodeposit Brinell number*	*Brinell hardness* Fully annealed	Work hardened
Chromium	Chromic Acid	400–950	70	
Platinum		606–642	47	97
Rhodium		594–641	101	
Palladium		190–196		
		385–435	49	109
Nickel	Sulfate	125–420	70	300
Nickel	Sulfate with Organic Colloids	up to 550		
Iron	Sulfate and Chloride	140-350	69	148
Copper	Acid Sulfate	40–62	40	102
	with Colloids	up to 130		
Copper	Cyanide	130–160		
Silver	Cyanide	60–79	25	68
Cadmium	Cyanide	12–22	20	34
Zinc	Sulfate	40–50	33–40	52
Tin	Stannate	8–9	4–5	

Hardness

The hardness of a deposit may be changed by changing the current density or temperature of a bath. The hardness is also greatly affected by the presence of organic substances, impurities and addition agents.

Hardness ranges shown in Table 3 were reported by P. J. Macnaughton and A. W. Hothersall.[3] The values in this table show that in plating it is possible to attain hardness values much greater than those obtained in work hardening.

Layered Deposits

Microscopic cross-section examination will often reveal layered deposits. These may be caused by addition agents, impurities, or metals that do not fit into the crystal lattice. A condition such as this was produced by addition of lead to a cyanide copper bath.[4] The lead will not fit into the copper lattice so that lead ions build up near the cathode; after sufficient copper has deposited to diminish its concentration near the cathode, the lead deposits in a layer. Another copper layer is deposited over the lead layer and the process is repeated. Small amounts of lead brighten the copper deposit, but large amounts cause an inferior plate.

REFERENCES

1. *Trans. Electrochem. Soc.*, **17,** 571 (1921).
2. *Trans. Faraday Soc.*, **31,** 1253 (1935).
3. P. J. Macnaughton and A. W. Hothersall, *Trans. Faraday Soc.*, **31,** 1168 (1935).
4. W. R. Meyer and Arthur Phillips, *Trans. Electrochem. Soc.*, **73,** 377 (1938).

4. PREPARATORY STEPS OF PLATING

Quality plating is primarily dependent on the reaction that takes place at the cathode during deposition. This reaction can only be favorable if the plating bath has been properly prepared and adjusted. But even if conditions are right in the plating bath and at the cathode, there is no guarantee that the deposit will be satisfactory. The quality of the deposit is also dependent on the condition of the cathode and on the preparatory steps. If the article to be plated is excessively porous, nothing can be done to produce a sound, continuous deposit, short of remelting the basis metal and making a new article. It is therefore necessary to make sure that the article to be plated is of good quality and properly prepared prior to plating.

The proper preparation for plating on item starts with the making of a fresh plating bath. If the bath is made up properly, it is possible to begin plating immediately without initial chemical analysis and without adjustments of the chemical content. However, many baths operate better if some initial electrolysis is carried out prior to plating of production pieces. An alkaline tin bath may be made up and used immediately, but a nickel bath almost always works better after initial electrolysis. Other baths, such as cyanide silver, require only filtration prior to use. Still others require filtration, electrolysis, and treatment with activated carbon or clay.

Preparation of the Bath

The first step in preparing a new bath consists of thorough cleaning of the empty plating tank. It is far easier to clean the tank when it is empty than to remove the dirt from the solution after the bath is made up. All solid particles and dirt should first be removed from the tank by sweeping and wiping. After the tank is thus swept it should be examined to determine if further cleaning is required. If the tank contains grease, this should be wiped out with a rag containing a suitable solvent. Further cleaning depends on the type of material of which the tank is constructed. Rubber or plastic tanks require no further cleaning, whereas wooden tanks may require painting or lining with pitch. If the tank is made of steel its surface will likely be covered with a visible oxide and possibly deposits of oxide and slag at the welds. Often the steel will be covered with a tight black oxide of iron that in itself is a good surface. It is difficult, however, to determine by examination whether the oxide will or will not be loosened after the bath is prepared. Therefore, it is best to remove the scale by pickling or sand blasting.

After the tank is cleaned and rinsed, it should be partially filled with water before the chemicals are added. If the tank is partially filled, it will be easier to agitate without spilling solution over the top. For very soluble salts, it will be convenient to fill the tank one-third, but for salts that are not very soluble, such as the total salts added to a high-concentration nickel bath, the tank should be filled about two thirds.

The order of mixing chemicals is important. If a cyanide copper bath is to be made up, the alkali cyanide must be added before the copper cyanide because the copper cyanide is insoluble in water. The order of addition of the carbonate, caustic, and Rochelle salt is relatively unimportant. Since the chemicals are not pure and since the tap water is hard, there will be a precipitation of salts in the freshly made bath. In many baths, these salts must be removed by filtration and they are more rapidly removed while the solution is concentrated. Thus, it is often time saving to start filtration while the solution is concentrated. The bath should then be made up to volume after most of the precipitate is removed since more salts may be precipitated on further addition of water.

At times, it is convenient to make concentrates for the purpose of preparing fresh baths and for additions to old baths. The advantages of concentrates are that they can be filtered and are more convenient to add to baths than salts in solid form. The disadvantages are that they require storage space and they increase the volume of the bath to a greater extent than dry salts (it may be necessary to remove a part of the bath).

When large volumes of solution are made, the salts should be added slowly and the solution agitated. Agitation may be provided by hand stirring with a large paddle, but if many solutions are made, a portable electric mixer will be convenient. If the salts are added too rapidly, they may form a cake on the bottom of the tank that will be very difficult to dissolve.

After all the salts are in solution, the specific gravity of the bath should be measured. If the right amounts of salts were used and they were all dissolved, the proper gravity will be obtained. If the gravity of a fresh solution is not known, it can be determined by making up 1 liter of solution and measuring the gravity. The gravity measurement is an immediate check on the over-all salt content.

After the bath is of suitable gravity and filtered, a plating test should be made. This may be done by either plating on a simple flat cathode, plating on an actual article, or (best) by running a plating range test. If plating-range standards are not available, it is well to record a sketch of the appearance of the plate in the log. The plating test will indicate whether the bath may be used as is. Adjustments of pH or addition agents may be required, or the bath may need initial electrolysis and purification.

It is often convenient to make a complete chemical analysis of a fresh bath, although this is not necessary if a gravity check and a plating test are made. However, a record of gravity, analysis, and plating test is valuable for the log until enough data are accumulated to dispense with some of the analyses.

Purification steps are best carried out simultaneously. These consist of filtration, carbon treatment, and electrolysis. The activated carbon should be loaded into the filter and the solution pumped over the carbon and through the filter. A large cathode may be suspended in the tank and the bath electrolyzed until the deposit looks good either on the cathode or by a plating test. Electrolysis for purification is most effective at low current density—5 amperes per square foot.

After the bath has been adjusted to chemical limits, purified and tested by plating, it should be tested at full load. The tank load can be calculated from the number of racks that it will hold and the current per rack. A cathode area of approximately full capacity should be used and the proper number of anodes should be placed in the tank. The number of anodes can be calculated from the total current, the area of the anodes, and the recommended anode-current density.

The tank is operated at full load in order to check the behavior of the current. The tank voltage is noted and the tank examined for electric faults and good contact of the racks and the anode hooks at the bus bars. If several tanks are supplied with current from one generator, it is well to check the effect of full load of one tank on the operation of another at full load. An electrolytic cleaner suddenly drawing current may cause considerable drop in tank current if the total current is near the capacity of the generator. This will only be serious if it interferes with plating thickness and the effect can be estimated from the observed current drop.

If a bath is being set up for the first time, it will be profitable to keep good records. If an adjustment of pH is required, record the amount of acid or alkali added and the pH before and after additions. From data of this sort, a graph can be constructed that will show the exact quantity of chemical to be added to bring the pH from any point to within the required pH limits. Useful data can be accumulated in a similar manner for adjustment of surface tension where this is required. If a few gravity readings are taken before and after chemical additions, a chart or graph can be prepared that will indicate the total salt for a gravity reading. A little extra time during the initial preparation of a solution will save time on all future adjustments. Charts and graphs prepared from initial data are always useful. When calculations are made for the amount of salt to be added to the bath, these should be recorded along with the tank volume.

After the bath is found suitable for full-scale operation, the geometry of the plating system should be considered. Anodes should be spaced so that the current reaches the racks equally from all directions, unless a special current-distribution problem exists.

The plating racks should be examined to determine whether the solution drains readily and is not trapped in the racks, causing drag-in and rinsing troubles. It is also well to check the rack contacts. This can be done by hanging pieces on the rack and checking with an ammeter across each contact for equal ability to carry current from the rack to each individual piece.

The efficiency of hoods, heaters, and means of agitation should be examined.

At the same time as the plating tank is set up, the rest of the plating line should be put into operation. This will include rinsing, cleaning, etching, and special steps such as striking or bright dipping.

Clean and Active Surfaces

The surface to be plated must be "clean and active" and free of grease, soil, and scale. Grease and soil is removed from the surface by solvent or alkaline cleaning. Scale is removed by chemical means such as pickling, or by mechanical means such as sand blasting.

The surface is made active by complete removal of all surface films. After it is rendered active it must be kept in this state until it is covered by the metal being deposited. Pickling will activate plain low carbon steels. A steel that has been so activated will remain active if it is immersed in an acid plating bath. Thus a pickled steel can properly be plated with nickel in an acid nickel bath. The nickel will then "bond" to the steel. The resulting bond will be stronger than the weaker of the two metals that form the bond.

If steel is cleaned, pickled, and then plated in an acid copper bath, the resulting deposit will not be bonded. The surface now is too active or too reactive to acid copper. Copper in the acid bath deposits chemically. In so doing steel is dissolved as copper is deposited and the resulting deposit is loose. This is immersion plating or plating by chemical displacement of the steel with copper. The problem is remedied by plating the activated steel with a copper strike from a cyanide copper bath. Acid copper can then successfully be plated over the cyanide copper. The pickled steel is properly active for copper striking and the cyanide copper plated surface is active for acid copper plating.

A striking step is an interesting variation of activation, worthy of detailed consideration. Consider that pickled steel can be kept active in a cyanide solution: (Cyanide holding solutions are common in plating shops.) Cyanide in a strike solution serves as an activating factor. The cyanide readily dissolves freshly precipitated iron hydroxide that forms when pickled surfaces are rinsed. Cyanide alone, however, is not enough to assure striking action. Nor is metal cyanide sufficient; a high efficiency copper cyanide or high efficiency silver cyanide bath will not act as a strike. The strike must be a low efficiency bath that preserves activation by the release of hydrogen while metal is being deposited. The strike requires cyanide, hydrogen release, and metal deposition—immediately and simultaneously. High efficiency copper or silver deposits can then be bonded to strike deposits.

Immersion plating frequently destroys bond, but sometimes promotes it and acts as a means of preserving activation. The best example of this is the zincate solution used to prepare aluminum for plating. Aluminum is cleaned and etched in acid, prior to zincating. It is then dipped in the highly alkaline sodium zincate solution. A very thin layer of zinc is immersion plated. A copper strike is usually applied over the thin zinc and many metals can then be deposited over the copper.

Metals can be activated by mechanical means. A vapor blast, consisting of a pressure blast with water and abrasive, will activate a surface. A dry blast will also activate a surface. These treatments are rarely used but are helpful if it is desired to restrict the amount of hydrogen released, such as when hydrogen embrittlement is a problem. The treatments however are followed by a short cold acid dip to assure activation.

General steps applicable to many plating cycles and that should be kept in mind are: (1) cleaning, (2) descaling, (3) activation, (4) plating.

Passivation should also be kept in mind. This is the formation of a film that acts as a barrier to bonding of electrodeposits. Steel is normally passive to all types of plating and cannot be successfully plated until it is activated. Once active it can readily become passive again. For example, steel will become passive in the few seconds that it takes to enter a strike bath. If the current is off as it enters, the plate will not be bonded. If the current is *on*, it will be bonded. If a pickled steel is allowed to remain in a rinse too long it will become passive—which accounts for the popularity of cyanide holding baths.

Stainless steels and nickel become passive much more readily than steel. These metals are rendered active by electropickling in an acid nickel bath. The current is then reversed and the parts are actively plated before removal from the same bath.

No phase of plating is more essential to success than a thorough respect for the principles of activity and passivity during preparation of the surface and plating of the first layer of metal.

5. PREPARATION OF THE SURFACE

We know that the surface must be cleaned of all foreign matter and that it must be chemically prepared for plating. The preparatory steps are dependent on the surface and on the plating bath. Furthermore, there are variations in the preparatory steps that are necessary to accomodate alloys, heat treatments, mechanical stresses in the metal, impurities, and contaminants worked into the surface. As a consequence there are many variations of a plating cycle, The plating cycle itself can be broadly defined in terms of limitations imposed by the basis metal. The preparation of steel, for example, is greatly different than the preparation of aluminum.

Low Carbon Steels

The following preparatory steps, with appropriate rinsing, apply to low carbon steel: (1) pre-clean, (2) clean, (3) pickle, (4) de-smut, (5) acid dip.

All oil, grease, and other organics are removed by the first two steps or an adequate single cleaning step. This cleaning removes the contaminants that could interfere with pickling. Step 1 utilizes solvent, emulsion or spray cleaning, or vapor degreasing, as necessary. Step 2 is an alkaline cleaning step—either electrolytic or immersion. After cleaning and rinsing, the part should be examined for absence of loose dirt and oil. If the part is not clean or shows water breaks it should be recleaned or perhaps submitted to more drastic cleaning action. The part is not necessarily clean just because it appears so—it can be contaminated with organics that are undetectable and also may sustain a water film. Usually, though, if the part appears clean at this point it has been satisfactorily treated.

Rust, oxides, or scale are removed by pickling in sulfuric or hydrochloric acid. If light scale or rust has readily been removed the part may be ready for rinsing and plating. Pickling should be kept to whatever minimum time will assure adhesion. When a single acid step is sufficient it is advisable to pickle the part until the part gasses freely, plus another half-minute.

After pickling, the part should be rinsed and examined for residual scale, dirt, smut, and water breaks. If any tight scale is present, further pickling is in order. Loose scale, dirt, smut, or water breaks call for a re-cleaning step. Anodic electrocleaning has been found beneficial at this point.

If the part appears clean after pickling it should be examined for "carbide smut" that comes from the steel. If this smut is light it is not readily detected: It is revealed by gently running a finger tip over the surface. When a light smut is present a gentle wipe will leave the area a little cleaner in contrast to adjacent areas. A very light smut that is hard to detect is not harmful to the work or the plating bath. Heavier smuts will be produced by higher carbon steels and by longer pickling times. These should be cleaned away by alkaline cleaning if possible or by power cleaning or even by swabbing of the surface. After smut is cleaned from the surface the parts should be activaged by a dip of 15 to 30 seconds in dilute acid at room temperature. It is not necessary for the parts to gas at this point. Very light gassing is permissible, but repickling again produces smut.

Parts that have been pickled can be rinsed and held in a cyanide solution. After a rinse, an acid dip, and another rinse they are ready for plating. In some cyanide baths the parts may be rinsed then plated. If the parts are to be held for long times they may be held in an alkaline solution, in which case they are rinsed and then lightly pickled prior to further processing.

Medium Carbon Steels

In order to discuss problems with higher carbon steels, this writer takes the liberty of defining "medium carbon steels" as 0.20 to 0.35% carbon. Above 0.35 the steels are arbitrarily defined as high carbon.

Medium carbon steels are more difficult to prepare than low carbon. The same methods should be tried with more attention to carbide smut. The objective—to activate the steel without forming excessive carbide on the surface. Even if the carbide is swabbed or cleaned away, the surface probably will not be receptive to a bonded deposit. Many of the steels contain molybdenum, chromium, vanadium, and tungsten that tend to pacify the surface. It is not possible to predict from the chemical composition and heat treatment the pickle treatment necessary. If a positive statement can be made about the steels in this class it is that "successful preparation is most likely with the less common methods." It is best to expect to change the preparatory procedure.

Of the unusual procedures, very few of them are successful in general, but a particular problem can be solved by one of the procedures, as follows:

1. Electroetch anodically in $FeCl_2$, 10 oz/gal; HCl, 1 oz/gal; c.d., 50 amp/sq ft; time, ½ min.
2. Rinse.
3. Anodic clean in an alkaline-phosphate cleaner one minute.
4. Dip in HCl 13 oz/gal for ¼ minute.
5. Rinse.

Similar anodic etching procedures using sulfuric, chromic, or oxalic acid solutions followed by an activation dip in hydrochloric acid have been found effective.

Chemical oxidation followed by a dip in hydrochloric acid was found to be successful with a number of steels:

1. Chemically oxidize in NaOH, 100 oz/gal; $NaNO_2$, 15 oz/gal; temp., 300-320°F; time, 15-30 min.
2. Rinse.
3. Dip in hydrochloric acid until the oxide is removed.

When this method is used the part can be rinsed and dried after oxidizing and then dipped at some later time. Thus, the chemical oxidizing can be done at an early convenience or in a remote area.

High Carbon Steels

A simple pickle will not suffice to prepare high carbon steels. A procedure must be used that will activate the surface and also will either avoid smut or rid the surface of smut prior to plating.

Some of the procedures applied to the medium carbon steels will apply to the high carbon steels but with less general success. Procedures applied to the stainless steels, particularly the activation-plating procedure (described later) are more successful.

Some of the high carbon steels are heat treated to quite high strength levels. At strength levels of 180,00 psi or greater (or Rockwell C 40 hardness or greater) they must be given special attention. At this strength and above, the steels are susceptible to cracking due to hydrogen embrittlement.

Severe hydrogen embrittlement has been known to cause a part to crack in the pickle. Better known failures due to this cause have occurred after parts have been placed in service. This is a problem that must be shared by the plater and by the designer if disastrous failures are to be avoided. Steels that are heat treated to Rockwell C 45 or greater are even more critical than those in the range C 40 to C 45. Hardened steels that have been embrittled with hydrogen will fail in time when placed under a sufficient load—one that is normally within design limits for an unembrittled steel.

In order for high strength steels to be unembrittled they must either be prepared without introducing hydrogen or else the hydrogen must be driven out after plating. In addition to these restrictions some steels are more susceptible than others.

Hydrogen embrittlement cannot be justly dealt with in a few paragraphs. When a problem exists or is suspected, an intensive study should be made or a reliable and qualified source should be sought to process parts. On the other hand, anyone who is associated with plating should have some knowledge of this potential hazard.

Hydrogen that is released into the steel during plating can be driven from the steel by heating after plating. Heating for several hours to 24 hours at 375°F is the usual treatment. This will drive hydrogen out of nickel or chromium plated steel. It will not drive hydrogen out of cadmium plated steel. Cadmium acts as a barrier and prevents the hydrogen from escaping. Special processing is required to cadmium-plate steel so that it is not embrittled.

There are no reliable general rules for proper plating of high strength steel to satisfy all designs. There are, on the other hand, many specific instances of the use of high-strength-plated parts.

The reliability of a high-strength-plated part is best established by testing for embrittlement. This is done by some means of submitting the part to a sustained load for a prolonged period of time. In many of the tests, a part or a standardized test specimen is plated by the practice in question and then submitted to a sustained load of 50 to 80% of the ultimate strength. The specimen is often notched to increase the sensitivity to failure. If it sustains the load for 100 to 200 hours without cracking or breaking, the process is judged satisfactory.

Hydrogen can be introduced into steel by any cleaning, pickling, etching activating, or corroding process that is cathodic, or by any plating process that is less than 100% efficient. The steel will tolerate some hydrogen, but failures have occured with the presence of only a few parts per million of hydrogen.

Mechanical descaling procedures avoid hydrogen release. Dry sand blasting is a useful method. Dry tumbling and grinding are applicable. Alkaline cleaning should either be soak or anodic. The most damaging treatment is an active or vigorous chemical pickle that etches the steel while releasing hydrogen. An anodic etch in cold acid or in an oxidizing acid is much less damaging.

Sand blasting or vapor blasting can serve as a substitute for pickling. It has been established that these treatments have an activating effect similar to a pickle. They should however be followed by a short dip in a dilute hydrochloric acid solution to free the surface of readily formed but easily removeable passive films. It need hardly be said that this acid dip must be kept very mild.

When light oxide or rust is present on high carbon steels it is best removed by pickling in hydrochloric acid. This will produce a smut and pacify the surface. The surface sometimes can be prepared for plating by anodic alkaline cleaning followed by anodic etching in sufuric acid.

Metal can be removed without smut formation by electropolishing in hot concentrated phosphoric-chromic acid solutions at high current densities. This usually leaves the steel in a passive condition. Anodic cleaning followed by anodic etching or acid dipping may activate the surface.

Unusual procedures must occasionally be developed to plate high carbon steels. A change in the type of steel or even a change in heat treatment with the same steel may call for a new process. A good appearance is no guarantee of success. The surface can be free of smut and have a light matte etched appearance yet not provide an acceptable surface. Conversely an irregular, splotchy surface may prove satisfactory. The surface must be free of contaminants and all but a light smut, but beyond this the appearance is not a criterion.

Stainless Steels

Stainless steels are difficult to plate because they easily become passive. It is this fact that makes them stainless steels. The problem is to activate the surface and to keep it active until plating has started.

Some stainless steels can be activated by cathodic treatment in sulfuric or hydrochloric acid or even by pickling in these acids. If steel remains passive in sulfuric acid, pickling action is started by touching it with a piece of non-stainless steel. Parts should be pickled for a minute after gassing starts. They should not be pickled longer because stainless steels also present a smut problem.

The most successful method for plating stainless steel is to activate it with an activation-plating bath. Indeed, this is a good general method for plating of any steel that is difficult to plate. The bath is a nickel chloride bath with enough hydrochloric acid present to cause activation: Ni $Cl_2.6H_2O$, 30-35 oz/gal; HCl, 5-10 oz/gal.

Nickel electrodes are used in the bath and the part is etched by anodic treatment one or two minutes at a current density of 20. The current is then reversed and the part is plated with nickel at a current density of 20 for 2 to 20 minutes. The principle of this process is that the part is activated in the bath and remains active until plating is started. A sufficient deposit of nickel protects the part from passivity and it is then easily rinsed and transferred to another bath for plating. Variations of this process are used and it has been found efficacious in plating nickel, cobalt, and other passive alloys. The cathode efficiency of this bath is low, therefore it builds up in nickel and uses acid. The bath also builds up in iron and other metals present in the alloy being activated, which then degrades the quality of the deposit. Portions of the bath must periodically be removed and discarded and acid added to maintain the quality.

Zinc-Base Alloys

The preparation of zinc like other metals responds to the cycle: (1) clean, (2) rinse, (3) activate, (4) rinse, (5) plate.

The thing that must be kept in mind with zinc is that the metal is quite reactive. It reacts readily with either alkaline or acid chemicals and easily replaces metals from metal salt solutions. Preliminary treatments must be with light or medium duty alkaline cleaners and dilute acids, or at lower temperatures and shorter periods of time. It is a temptation to provide general cleaning and pickling with a minimum number of solutions—for instance, by using the same cleaners and pickles for both zinc and steel. It may be possible to use the same solutions for both but it is quite likely that such solutions would be compromised to the point where they would be not quite right for either. Compared with steel the treatment of zinc must be gentle.

The preparation of zinc-base die castings really begins with polishing and buffing. Polishing is done with 180 to 220 grit abrasives to smooth rough areas and prepare for buffing. Buffing is done in one or more steps on cloth wheels with the use of buffing compounds. An effort should be made to remove the "polished" finish and then final buff so that the final "color" is produced with a minimum of compound left on the work. Compound that can be removed by dry buffing will simplify cleaning.

Compound that remains on the work should be removed by precleaning and if it is caked it should be removed before it hardens further. It is best to remove or at least loosen the compound by vapor degreasing or emulsion cleaning.

Two-stage alkaline cleaning may be beneficial in several instances. First, it may aid the removal of heavy compounds that were only partially removed by solvent precleaning steps. Second, the first alkaline cleaner may be the most economical means of precleaning to provide quality cleaning at the second stage.

Single-stage alkaline cleaning or final alkaline cleaning should only apply to a surface that is relatively clean, requiring only the removal of a light oily film.

Prolonged alkaline cleaning, whether it be in one or two stages, is undesirable. When prolonged cleaning is necessary it is because soil clings to some areas for prolonged periods. This means that one area is cleaned more readily and is exposed to attack for longer times than other areas resulting in uneven attack and spotty work.

Final cleaning is frequently done by electrocleaning, usually anodic. Spray cleaning is also used, with the advantage that it combines flushing with alkaline action when needed.

The alkaline cleaners attack the zinc and leave an alkaline film on the work that must be removed and neutralized in the acid dip. Also, a very mild acid attack of the metal surface is desirable. This is done by dipping in dilute acid for about a minute. Acid solutions are ½ to 1 oz/gal. Sulfuric acid is usually satisfactory, although hydrochloric, hydrofluoric, citric, and fluoboric acids have been found beneficial.

Generally, the zinc is plated in a solution that readily attacks the metal. In order to avoid this attack, the zinc is struck with copper. This avoids etching, immersion plating, and loss of bond in a following nickel plating solution. The most general plating cycle then is: (1) pre-clean, (2) rinse, (3) clean (4) rinse, (5) acid dip, (6) rinse, (7) strike, (8) rinse, (9) nickel plate, (10) final plate as desired.

The key step is the strike consisting of a warm Rochelle salt cyanide copper bath at a pH of 11.2 to 12.5 applied in ½ to 2 minutes. This prepares the bath for heavier copper plating, or copper then nickel, and finally chromium or any other desired plate.

Aluminum

Aluminum has many of the characteristics of zinc but it is much more active chemically. In order to plate on aluminum a precise preparatory cycle must be established that will treat the alloy gently and provide an intermediate film that will protect the substrate while the first plating is taking place. This can be done in several ways: by anodizing prior to plating or by depositing an immersion coating to which the plating can be bonded. The most common commercial means is the zincate method, which chemically deposits a thin film of zinc from an alkaline zincate solution. Sometimes this is done with a single zincating step, but mostly it is done by zincating, stripping, and rezincating—known as the "double zincate." This double process is generally recommended.

Rigorous compliance with established practice starts with cleaning and etching, followed by zincating and finally plating. The surface must be cleaned and activated and then zincated to protect the activated surface. The subsequent plate is applied over the zinc, replacing it in part as deposition starts. Chemical activity continues on the surface as the first metal is deposited.

There are many established variations of the zincating process but there is no single process to plate all alloys. Some experimentation should be expected to establish procedures. One might wonder why this should be so. It is because of the fact that minor changes can make a difference in this sensitive process. Such things as differences in the water used, rinsing practices, the rate of transfer from one tank to another, and particularly the condition of the surface can make a distinct difference in this process. Close attention to detail and reproducibility is mandatory to maintain a reliable procedure.

A simple zincate cycle that sometimes works consists of: (1) clean, (2) etch, (3) zincate, (4) plate.

The double zincate cycle is more elaborate: (1) clean, (2) etch, (3) de-smut, (4) acid dip, (5) zincate, (6) acid dip, (7) zincate, (8) copper strike, (9) plate.

Deviations from the simple cycle and the double cycle vary in complexity between the two.

Prior to etching it is of course necessary to remove oil and grease from the surface. This is done by conventional cleaning including pre-cleaning where required. Inhibited cleaners are usually used but etching cleaners also have a place. When an etching cleaner is used it will not substitute for the acid etch that has been found essential prior to zincating.

If the aluminum is etch-cleaned it will be necessary to de-smut. This step can at times be done in an acid dip that will serve to de-smut and activate.

There are no visual criteria that assure success. A good clean appearance after etching and a uniform gray coating after zincating esthetically suggest success, but too frequently have been completely unsatisfactory. Results have been satisfactory with a mottled unsightly appearance; mere repetition of this appearance is not a dependable criterion of adherent, acceptable plating.

Through the cleaning and etching steps the process is not critical; these can be done and reproduced quite easily.

Zincating is critical. It can be overdone and the film can be damaged in the subsequent rinsing step or in the strike before sufficient metal is deposited to protect the aluminum.

The zincate film is easily dissolved in an acid dip. Removal of the film and rezincating comprises the double-zinc method. The second zincate produces a different film than the first zincate. When this process is used, the first zincate is only a part of the activating cycle prior to applying the final zinc film over which a strike or other deposit will be applied.

After the zinc immersion layer is applied it is possible to deposit metals directly from solutions that do not readily attack zinc. Most of the plating solutions are sufficiently acid or alkaline that they attack zinc and require a copper strike over the zinc prior to plating, or a copper strike plus a thin copper deposit.

The simple cycle is preferred when it will work. It is most likely to succeed with alloys that are low in alloying elements—1100 and 3003 alloys. A simple dip in 1:1 nitric acid can prepare the metal for zincating.

Nitric acid is almost unreactive with aluminum and is usually used as a de-smutter. It has, however, been found sufficiently active for some zincating processes. With 5000 and 6000 series alloys a more active etch such as nitric-hydrofluoric or chromic-sulfuric-hydrofluoric become essential. A dip for 15 seconds in an etch within the following range has been found to suffice: CrO_3, 8-16 oz/gal; H_2SO_4, 15-30 oz/gal; HF, 2-4 oz/gal.

Acid dips containing hydrofluoric acid also are recommended for treatment of the cast alloys (those that contain silicon).

The double zincate process is recommended for alloys containing copper—2000 and 7000 series. Also, it is generally recommended when the simple cycle is not sufficiently reliable. Etching prior to double zincating is not as critical with the double cycle. The first zincate followed by a dip in nitric acid provides proper etching-activation. In some instances, this will suffice so that etching prior to the first zincate may be eliminated.

A popular zincate bath employs a concentrated solution: NaOH, 55-65 oz/gal; ZnO, 10-13 oz/gal.

Drag-in of water from the preceding rinse will eventually dilute this bath below the operable range, and frequent specific gravity checks will reveal changes due to this. One concentrated bath was maintained at a specific gravity of 1.404 by additions of sodium hydroxide and zinc oxide in the proper ratio.

The viscous solution presents a drag-in problem in the rinse that follows. In this rinse, the clinging zincate solution should be rapidly moved away from the surface to avoid alkaline attack. Heating of the rinse to 120°F or sufficient agitation will accomplish the desired movement of the diluting zincate from the vicinity of the surface.

A Rochelle copper cyanide strike bath operated at room temperature will suffice for copper striking: copper cyanide, 5.0-5.5 oz/gal; free sodium cyanide, 0.3-0.5 oz/gal; sodium carbonate, 2.0-6.0 oz/gal; Rochelle salt, 6.0-8.0 oz-gal.

The pH of the strike may rise because of drag-in from the previous alkaline rinse. It should be kept in the range of 11.0 to 11.5. Striking for several minutes at 20 amp/sq ft should be satisfactory.

Copper Alloys

Copper surfaces that are freed of dirt, scale, and stain will be active.

The surface of the metal will not readily become passive as with the less noble metals. Because of this, copper is relatively easy to prepare for plating. On the other hand, copper alloys stain readily and this must be avoided just prior to plating.

Conventional pickling, descaling, abrasive cleaning, and precleaning provide preliminary preparation to remove scale, stain, buffing compounds, and organic materials.

When light contamination and stain are present, a typical cycle is as follows: (1) electroclean, (2) acid or cyanide dip, (3) plate.

Electrocleaning is not essential but it is recommended. Anodic cleaning promotes light staining, whereas cathodic cleaning produces a very light smut. Anodic cleaning for a few minutes or cathodic cleaning for 1 to 2 minutes followed by a 5 to 10 second anodic clean will produce a desirable surface. A light stain that is produced by anodic cleaning is beneficial. The stain is easily removed by a cyanide dip or a mild acid dip and it provides a simple activating procedure with a minimum of metal removed.

A cyanide dip in 4 to 6 oz/gal of NaCN is recommended prior to alkaline or cyanide plating. An acid dip in dilute sulfuric, hydrochloric, nitric, or fluoboric acid of 4 to 6 oz/gal is desirable prior to acid plating.

Rinsing steps must be inserted after cleaning and after dipping. Rinsing is somewhat critical. If the rinse becomes contaminated it will promote staining and cause plating difficulties. If the rinse builds up in copper and this is dragged into the plating bath it will cause noble metal contamination. Double rinsing or an adequate flow in a single rinse is important to avoid these troubles.

A cyanide dip followed by a very light pickle is sometimes employed.

A copper strike has often been found essential for plating of leaded alloys or soldered parts. If the work is to be plated in an acid solution the strike is followed by an acid dip.

A cycle (including the rinse) that is not uncommon is as follows: (1) preclean, (2) rinse, (3) alkaline clean, (4) rinse, (5) cyanide dip, (6) rinse, (7) strike, (8) rinse, (9) acid dip, (10) rinse, (11) plate (acid).

Although this cycle is somewhat elaborate it is relatively gentle and is recommended for buffed work. When it is not necessary to retain a polished appearance it is more expeditious to use a mild pickle in place of an acid dip. Slight etching in nitric-sulfuric or ammonium persulfate will activate the work and eliminate the strike.

Surface appearance of copper alloys is enhanced by bright dipping in concentrate sulfuric-nitric-hydrochloric bright dips. The bright dip can serve as an activator and become a part of the preparatory cycle.

Preliminary treatments must be modified to suit the alloy and the heat treatment. Beryllium copper in particular will not respond to a simple treatment. Oxidizing pickles followed by a bright dip are used to prepare beryllium copper for introduction into a conventional cleaning-acid dipping-striking cycle.

Other Metals

Magnesium is an active metal roughly similar to aluminum. It is not surprising therefore that one of the most promising methods of plating magnesium has been the zincate method—clean, activate, zincate, strike, plate. The solutions used and the process details are, however, different from those used for aluminum.

Heat resistant alloys of nickel and cobalt have successfully been plated by the nickel activation-plating methods used to plate stainless steels.

Titanium, molybdenum, tungsten, and less common metals have been plated. Eventually, someone wants to plate every substrate. There is a wealth of experience that has been accumulated and recorded to serve as a guide. On the other hand, each application is a little different than what has gone on before—enough different to require experimentation. Where no information is available it is assumed that a method can be found. This can be justified from a basic observation that comes from electroplating experience—that electroplates can be bonded to metallic substrates.

6. CLEANING

As stated on p. 28, the surface to be plated must be clean and active. It must be free of all substances that will interfere with the bond, the uniformity, or the adhesion of the deposit. This will include removal of all superficial substances such as grease, soil, dirt, buffing compounds, metal chips, and loose scale. These substances can be removed from the surface by treatment with solvents and alkaline cleaners. Such treatments leave the surface clean without removing metal, but also without removing tight scale, rust, or inert substances that are bonded or embedded in the surface. Removal of tight substances may be regarded as a part of the cleaning cycle although it is more generally assigned to the pickling step. In fact it is well to regard pickling and cleaning as complementary steps. Cleaning prepares the surface to be pickled. In order to pickle, the metal must be sufficiently clean that the pickling solution will wet, contact, and react with the entire surface.

Scale and rust is removed by "descaling." This can be done by abrasive cleaning or by pickling. Descaling may then be a step preceding pickling or it may be a part of the pickling step. Pickling may be preceded or followed by cleaning. Pickling to remove heavy scale may require excessive treatment that will leave a smut on the surface. Such a smut must be cleaned away prior to plating—by vapor blasting, alkaline cleaning, or chemical de-smutting. After the final cleaning the part is then dipped to activate the surface (usually with an acid).

Loose scale will often trap grease that is not readily cleanable. This is overcome by cleaning, pickling to expose the grease, recleaning, then repickling.

At times, steps are combined. A common example is etch-cleaning of aluminum by the use of an alkaline cleaner that removes grease and soil while simultaneously etching the aluminum.

Plating, or at least pre-plating, can be combined with cleaning or pickling, although this is rare. In one instance, a small amount of sodium stannate was added to a cathodic cleaner so that a very small amount of tin was deposited during cleaning. This aided the removal of oil in recessed areas prior to subsequent flash alkaline electrotinning. The same thing could have been done by solvent cleaning prior to alkaline cleaning, but at greater expense. A better example of combined steps is the historical Bullard-Dunn process. This is an acid electrolytic process used to minimize pitting of steel during descaling. The work is made the cathode in a warm solution of sulfuric acid and tin sulfate. After treatment the work is left free of scale and with a thin film of tin. The tin can then be removed by anodic treatment in an alkaline cleaner.

Water-Break-Free Surfaces

Cleaning, in a narrow sense, only has to be sufficient to assure the success of the following step.

A milestone on the road to the clean and active surface is the "water-break-free" surface. A frequent inspection requirement is that a surface be clean to a water-break-free condition. This assures that the surface will be sufficiently free of organic films that it will "wet" with water—that the part will sustain a film of water when withdrawn from the rinse. A more rigorous requirement is that the part remain water-break-free after a rinse, an acid dip, and another rinse. Water breaks sometimes form on acidified surfaces that will not form on alkaline surfaces. Alkaline cleaning or abrasive cleaning are required to produce a water-break-free surface. These are good practical rules to which there are not too many exceptions.

Degreasing

In practical terms, water-break-free defines an organic-free or grease-free metal surface. The cleaning process is one that will remove grease and is sometimes called "degreasing." In the plating business, however, it is preferred to reserve the term degreasing for solvent degreasing. Strangely enough this degreasing does not assure a water-break-free surface. Vapor degreasing with a clorinated solvent removes practically all grease from a surface but leaves a non-wettable condition. Vapor degreasing followed by alkaline cleaning is a sequence that will almost always assure reliable cleaning. Alkaline cleaning is often sufficient in itself, but degreasing prior to alkaline cleaning adds reliability.

Pre-Cleaning

The need for degreasing points up the fact that complete removal of grease often demands a two step procedure. The first step consists of pre-cleaning—to remove excess grease or superficial organics that are too difficult for an alkaline cleaner. Pre-cleaning also keeps the 'cleaner' clean, so that objectionable contaminants will not be redeposited as parts are withdrawn from the cleaner.

Pre-cleaning can be done in many ways. Any economical means that will fulfill the requirement of preliminary cleaning is acceptable. Tumble cleaning, scouring, and spray washing are used. Commoner methods are vapor degreasing or emulsion cleaning.

Solvent Cleaning

Cleaning with cold solvents is convenient. There is always some type of solvent available—carbon tetrachloride, methyl ethyl ketone, kerosene, naphtha, alcohol, paint thinner, or trichloroethylene. It is inevitable that these be used for pre-cleaning of small parts and spot cleaning as well as removal of ink stampings, gum from tapes, paint residues, preservative oils, and lubricants.

Cold solvent cleaning is not much used on a large scale because it soon becomes contaminated, leaving films on the work. It then presents the expense of disposal or recovery for further use. Many solvents are a health hazard and so pose a drying problem. Hot vapor degreasing overcomes these major objections.

Vapor Degreasing

A vapor degreaser consists of a tank designed to maintain a volume of hot vapor. This is easily done by placing a heater at the bottom of a container and a cooling coil at the top. Solvent within the container continuously boils and condenses. The space between the boiling and condensing zones contains the hot vapor. A variation of this simple device contains an unheated overflow section. Condensed vapor collects in this section and keeps it full, warm, and overflowing—overflowing it to the boiling zone.

Some degreasers are equipped with a pump and hand-directed spray head. Clean condensed liquid provides solvent for the spray. An equipped vapor degreaser is used in many combinations of cleaning by dip, spray, and condensation.

Most cleaning in a vapor degreaser is done by merely allowing the hot vapors to condense on the cold work. Condensation and efficient flushing of all surfaces continues until the work reaches the temperature of the vapor. The method is unique among cleaning systems in that the cleaning medium, presented to the work, is continuously free of contamination. It is an efficient means of removing solvent-soluble organics from the surface. Also the part comes dry from the degreaser. It is a one-step cleaning and drying process.

Parts that are heavily covered with grease or that hold slowly soluble organics require more than immersion in a vapor. Vapor immersion is then supplemented by liquid immersion, in the overflow zone, or by spraying. The part can be soaked in the warm overflow to loosen and dissolve stubborn organics. Since the overflow liquid is cooler than the vapor, the part can now be moved to the vapor zone where condensation flushes away contaminated solvent remaining on the surface. Movement sometimes aids solvency and spraying then helps, particularly on intricate shapes. Spraying also helps to move inert materials that are held to the surface by the organic to be removed. Heavily contaminated parts can be placed directly in the boiling liquid.

A word of caution is in order at this point. A vapor degreaser is convenient to remove heavy layers of grease but it is usually not recommended to be so used. The reason is that the degreaser soon becomes loaded with grease, and expensive and time consuming cleaning and solvent recovery is then required.

Action in the vapor zone is limited by the heat capacity of the part. Parts that are removed while they are still being heated by the condensing liquid will carry out wet hot solvent. Parts must therefore remain in the vapor zone until they reach the vapor temperature. Massive parts may consume considerable time. This time can be reduced by immersion in the boiling zone, followed by flushing if necessary and finally condensation flushing in the vapor zone.

Tiny parts of low heat capacity will heat too rapidly to condense enough liquid to flush clean the parts. These may be cleaned by repeated withdrawal and reimmersion. Or better, they may be cleaned by placement under a condenser designed to produce a continuous flow of clean solvent within the vapor zone.

Two popular solvents used in vapor degreasers are trichloroethylene and perchloroethylene. They are preferred because of low toxicity, as compared to other chlorinated solvents. The solvents must be stabilized against decomposition by the use of "stabilizers." Stabilization avoids or minimizes the formation of acid (due to oxidation and hydrolysis).

Contamination in the boiling zone can be allowed to build up to 25% by volume (as oil). The solvent should then be boiled off to an increase of 50% by volume of oil in the residue. The residue is then removed and disposed of. This method utilizes the degreaser to recover solvent. The oil content of the solvent is easily determined by a gravity measurement.

The vapor degreaser is an efficient, versatile, unique cleaning device. It is the best means of doing many jobs. On the other hand the cost of the solvent is a distinct, limiting factor. Vapor degreasing does consume solvent—by loss directly to the atmosphere or to the ventilation system.

Emulsion Cleaning

Emulsion cleaners will preclean heavily contaminated surfaces. These cleaners provide solvent cleaning in a water medium. Hydrocarbon solvents are dispersed as a "milk" by means of surface active emulsifying agents. Dispersions of 1 or 2% allow organic cleaning will a relatively small volume of solvent. The presence of water aids removal of solid particles as the solvent removes or softens greases. It also sweeps away grease-laden solvent to allow fresher solvent to come into action. Temperatures in the range of 140 to 180°F greatly aid cleaning. Agitation also is beneficial and pressure spray emulsion cleaning is a very effective way of increasing the cleaning rate.

The work leaves the cleaner with a greasy film that can be reduced but not eliminated in a following rinse.

There are many interesting variations of this type of cleaning. In one, the work is soaked in an emulsifiable concentrate to allow maximum solvent action, and the concentrated residue is then emusified away by spray rinsing. This procedure is expensive in concentrate.

A compromise of the above two methods is accomplished by "di-phase" cleaning that consists of a solvent layer floating on water. This allows larger volume action in the solvent layer or even in droplets when it is used as a spray.

A further compromise between emulsion and di-phase cleaning uses an emulsion and a layer. This is accomplished by the use of volumes of emulsifiable concentrate up to 10% to produce an unstable condition or by the use of less stable emulsions.

The greasy film that remains after cleaning is put to beneficial use to protect temporarily machined parts against rusting or staining. Alternately, parts may be dipped in concentrate, which is allowed to remain on the work as a protective layer and is then rinsed away some days later. When these variations are considered it becomes apparent that this general water-solvent type of cleaning provides many possibilities to meet scheduled work flow and temporary storage in-process.

Alkaline Cleaning

An alkaline cleaner is expected to penetrate dirt, wet the surface, float-off grease, and leave a clean water-break-free surface.

An alkaline cleaner will remove greases and oils. After these are removed the dirt will be freed from the sticky grease and will fall away. Although this is an oversimplification it does pinpoint the primary essential action—namely, alkaline degreasing.

Degreasing by alkaline action takes place by emulsification, saponification, and dispersion. Strong alkalies will saponify animal and vegetable oils to form water soluble soaps. Soaps and other wetting agents promote emulsification of petroleum oils by lowering of the surface tension. This leads to the likely conclusion that good cleaning is possible with a solution of sodium hydroxide and soap. It turns out that this is not so—saponification and emulsification are of secondary importance. Cleaning is more dependent on surface active materials that surround the oil, float it off, and disperse it. In general, "surface active agents" includes soaps, wetting agents, and other alkalies and chemicals that promote separation of grease and oil from a surface.

Cleaner Testing

A great deal has been learned about cleaner formulation by the simple expedient of testing. Good cleaners can be separated from poor ones by observing the soak cleaning time to remove applied grease. A heavy paste made of petrolatum and rouge serves as a test grease. Cleaning away of a spread of this grease can be observed through a glass container. Another good testing layer is a baked-on oil. One part of heavy motor oil and one part of naphtha provides a test liquid. Test pieces are dipped, drained, and then baked at 350°F. A good soak cleaner will remove this baked film in 10 minutes. Immersion cleaner testing of panels prepared in these two ways has revealed the relative detergency of basic cleaner chemicals. For instance, it has been confirmed that a solution of 4 oz/gal of sodium orthosilicate at 180°F provides very good cleaning action. Sodium metasilicate was found to be a fairly good cleaner, trisodium phosphate only fair, sodium hydroxide poor, and sodium carbonate poor.

Heavy Duty Cleaners

Sodium metasilicate plus sodium hydroxide proves to be a good cleaner. Cleaner No. 1 of Table 4 is an example of a heavy duty alkaline-silicate cleaner. Alkaline-silicates disperse oils and greases, emulsify oils, and suspend dirt and oil. The cleaning action is aided by the presence of a small amount of wetting agent. Modifications of this formula account for a large portion of the heavy duty cleaners. These cleaners have earned the "heavy duty" title because they will remove heavy, tenacious, or stubborn soils that cannot readily be removed by a milder cleaner.

Plating troubles have been traced to silicate drag-out from silicated

Table 4 Alkaline Cleaners (oz/gal)

	1	2	3	4	5	6	7	8
Sodium hydroxide	4	2	2		1	1		
Sodium metasilicate	4	2	3	4				
Trisodium phosphate		2		2	4		4	2
Sodium tripolyphosphate			1					
Sodium hexametaphosphate						2		
Sodium carbonate		2	2		2	3	2	2
Naccanol NR	0.04							
Soap (soda)		0.08						
Soap (rosin)			0.3					
Wetting agent				as required (columns 4–8)				

cleaners. It is not common but it does point up that an inorganic film remains that is not rinsed away. If this film survives the intermediate wet steps it can promote rough plating. It can do this directly by remaining on the work or indirectly by contaminating the plating bath. It is a possibility to keep in mind in the event of trouble.

The heavy duty cleaners are used principally to clean steel and, to a lesser extent, copper and magnesium.

The heavy duty formula is converted to lighter duty by reduction of the highly alkaline sodium hydroxide and substitution of milder alkalis such as sodium phosphate and sodium carbonate. Sodium phosphate has some ability to disperse oils, although it is much less efficient than silicate. Phosphates and carbonates also serve as water softeners.

Multipurpose Cleaner

Experience indicates that some soils will respond to detergent action by the milder alkalis. Strong cleaners are at times objectionable because they promote staining of copper alloys.

Cleaner No. 2 is a medium duty cleaner that retains some of the good cleaning characteristics of alkaline-silicates modified by a more mild action—less tendency to film formation and some improvement in rinseability. Similar compositions are used to clean steel, magnesium, copper, and aluminum. It is good economy to use a cleaner that will remove many kinds of soil from many metals.

Surface-Active Agents

Occasionally, the success of a cleaner will be more dependent on surface-active agents than on a choice of the alkalis. Cleaner No. 3—a general purpose, medium duty, soak cleaner—might be such. This medium duty alkaline-silicated cleaner is modified by sodium tripolyphosphate, a good water softener that also enhances the action of some wetting agents. The wetting agent in this cleaner is a rosin soap. In each of the first three cleaners the wetting agent is different. These wetting agents are not partial to these particular cleaners: Various soaps and synthetic detergents may be used and indeed must often be used to assure efficient cleaning.

Wetting agents promote a foam blanket on cleaners, and foam control is important. A layer of foam reduces evaporation and acts as an exit blanket through which the work is drawn. The foam will practically eliminate obnoxious alkaline spray that is generated in electrocleaners. Some foams will trap dispersed oil in the foam in such a manner that the oil carried out with the foam, on the work, is easily rinsed away. Other foams allow oil to collect on the surface of the solution as a layer, where it is picked up to recontaminate the work in large non-rinseable spots. Wetting agents that form stable foams are not beneficial if the cleaner is vigorously agitated, such as with electrocleaners. Stable foams overflow the tank and are sucked into the ventilation system.

Silicate-Phosphate

Cleaner No. 4 is a strongly silicated cleaner modified with phosphate. It is a medium duty cleaner useful for soak cleaning of aluminum and electrocleaning of other metals. It resembles No. 1 in that it requires no initial carbonate. Carbonate is always present to a small extent as an impurity in the strong alkalis, in addition to which it continues to build up by absorption of CO_2 from the air. It could be omited from any of the first four cleaners. On the other hand it is of slight benefit as a water softener, and since it will build up anyway, it is usually added.

Light Duty Cleaners

Carbonate is recommended as an addition to light duty cleaners 5 to 8. Carbonate in these cleaners becomes important to alkalinity control because the cleaners are dominated by the milder alkalis.

Cleaner 5 is a good free rinsing electrocleaner that was developed as a final cleaner for machined steel that had been precleaned in a vapor degreaser. Cleaner 6 can also be used for this purpose but is more dependent on good precleaning than No. 5.

Cleaner 7 is a light duty electrocleaner used to cathodically clean steel, copper, tin, lead, and zinc.

Cleaner 8 is an alkaline etch cleaner used for etch cleaning of aluminum at 140 to 160°F. An etch cleaner is one that cleans by chemical alkaline etching plus simultaneous surface and detergent action. A similar result is frequently better accomplished by cleaning in a non-etching silicated cleaner followed by etching in a hot 5% sodium hydroxide solution.

Cleaner Formulation

A satisfactory cleaner is very often confirmed by the most direct approach—simply by cleaning production parts. This is relatively easy when parts are small enough to be cleaned in a trial 1-5 gallon bath. A great many practical cleaner tests can be accomplished on a small scale. Such tests, with baths made from known chemicals, can reveal the importance of each chemical that contributes to the composition. It will also reveal how a basic bath may be modified.

Consider the following general purpose medium duty cleaner:

Table 5 Medium Duty Cleaner

Chemical	*Composition % by weight*
Sodium hydroxide	25
Sodium metasilicate	40
Sodium tripolyphosphate	10
Sodium carbonate	24½
Wetting agent	½

This cleaner can be used at various concentrations and conditions as follows:

Table 6 Variations of Cleaner

Type of cleaning	*Concentration, oz/gal*	*Wetting agent*	*Temperature, °F*	*Time, min.*	*Current density, asf*
Soak	7–10	soap	180–boiling	5–20	
Spray	½–1½	synthetic	130–170	¼–1	
Electro	4–6	synthetic	160–180	½–5	20–60

Changing the concentration of the cleaner as a whole in effect varies the formula. This cleaner at 10 oz/gal approaches a heavy duty cleaner and at 4 oz/gal begins to resemble a light duty cleaner. However the chemical and surface activity of the cleaner is much better changed by changing the ratio of the chemicals in the bath. Sodium hydroxide can be increased in this cleaner to convert it to heavy duty or eliminated to serve for light duty. Other phosphates may be substituted for polyphosphate to favor cleaner stability at higher temperatures. As mentioned on p.50, very substantial changes can be made and indeed often must be made in the choice of the wetting agent.

Two-Stage Cleaning

It is, of course, desirable to clean in a single step. Often, it is more economical and more expedient to clean in two stages. The first stage is usually the precleaning stage with solvent, solvent vapor, or emulsion cleaner. This can also be done with an alkaline cleaner. A heavy duty soak cleaner can be used as a first cleaning stage. This is done to clean away soil that is difficult to remove or that will contaminate the cleaner. A part cannot be rinsed clean when drag-out of soil is excessive from a heavily contaminated cleaner. A second light duty cleaner or electrocleaner will remove this soil. This procedure is also helpful to remove films left by silicated cleaners.

First stage cleaning can be done early in the manufacturing process. When it is economical to clean twice it is sometimes better to clean first after a drawing or forming and prior to a machining operation. Second cleaning is then done just prior to plating. Cleaning of oils from the surface of a part just prior to heat treatment often solves a very difficult cleaning problem.

Planned Cleaning

The manufacturing department that is responsible for drawing or forming will use compounds that best aid these operations. Some of these, naturally, are more difficult to remove than others. Oils and fats are easy to remove, but soaps, solids, and additives are difficult, particularly if they cake on standing. Heat from forming will polymerize oils and burn soaps. Difficult films are similarly produced during buffing and polishing operations. Early cleaning aids the removal of these difficult films. When there is a choice of forming or buffing compounds it is fruitful to conduct tests to determine the cleanability of these. The same applies to rolling oils and quenching oils.

Soak Cleaning

It is convenient to clean parts merely by hanging them in a tank. This procedure is simple and equipment costs are low. Requirements are a heated steel tank, an exhaust system, means of racking, and a rinse. Cleaning is relatively slow, so it is helpful to clean at or near the boiling point. Boiling will reduce cleaning time by increasing the reaction rate and by providing agitation from the boiling action. Placement of heaters along one wall of the tank will aid thermal agitation. Problems with foam and cleaner spray are not as severe as with electrolytic and spray cleaners. If cleaning time is not a prime factor, this is a good method—frequently preferred.

Cleaning near the boiling point is not advised if lower temperatures will suffice. High temperatures increase the rate of carbonate formation, put a heavier load on the exhaust system, and cause trouble due to rapid drying of the cleaner solution before the part can be transferred to a rinse. Higher temperatures also promote staining of copper alloys. If parts are relatively free of soil or have been precleaned they can usually be cleaned by immersion at temperatures of 140 to 180°F.

Selection of an effective cleaner or change in the formulation to combat a specific soil is more often necessary with soak cleaners than with other types. A knowledge of whether or not the soils are emulsifiable, saponifiable, or dispersible can be helpful as a guide to formula changes.

Electrocleaning

Electrocleaning appears to be dramatic as compared to soak cleaning. Baked-on engine deposits, (on pistons and other parts for example) that cannot be removed by hours of soak cleaning can be removed in one or two minutes in an electrocleaner. In general, however, electrocleaning merely does the job faster, which is important when large numbers of parts are processed and often essential to automatic processing. Electrocleaning times are in the range of ¼ to 5 minutes, whereas soak cleaning times are 1 to 60 minutes. The efficient electrolytic agitation of these cleaners allows for reduction of other requirements. The cleaners can be operated at lower temperatures and with milder alkalis. This reduces staining and film formation.

If cleaning is difficult, it may be necessary to use a heavy duty electrocleaner. This is to say that stubborn soils can require the combined cleaning abilities of soak cleaning and electrocleaning.

Electrocleaning depends on the release of hydrogen or oxygen on the work when it is made a cathode or an anode. In order to be effective, the parts must be racked with sufficient spacing so that the current reaches all areas. The cleaning rate is increased by increasing the current density. Cleaning is slow at current densities below 10 amperes per square foot. The rate increases markedly as the current density is increased up to 40. Above 40 there is some further gain up to 70, but above 70 the gain is nil. A current density range of 30 to 50 is recommended.

Cathodic cleaning or "direct" cleaning releases hydrogen on the work; anodic "reverse" cleaning releases oxygen. At a given current density, twice the gas is evolved with cathodic cleaning as with anodic; however, experience shows that the cleaning rate is the same.

Cathodic cleaning of steel should be avoided when the cleaner is used in an environment or in a manner that submits it to contamination with plateable alkaline soluble metals—specifically lead, tin, and zinc. Anodic cleaning will promote solution of these same metals, so cathodic cleaning is preferred to avoid dissolution when the metals themselves are cleaned. Cathodic cleaning keeps these metals free of anodic staining. Chemical staining is still possible, however, and mild electrocleaners at moderate temperatures are used to avoid this type of staining on brass aluminum and zinc.

Cathodic cleaning should not be used to clean high strength steels that are subject to hydrogen embrittlement.

Spray Cleaning

Spray cleaning is a high production technique applied to parts that are of a favorable shape. The parts are conveyed through a spray cleaning machine on racks, link conveyors, baskets, or by a tumbling action. The method of conveyance and the part shape must favor impingement by the sprays on all surfaces.

The spray cleaners are of low concentration and depend greatly on the high pressure of the sprays for cleaning action. Wetting agents and soaps formed as a result of alkaline reaction with the soils will promote troublesome foams that must be controlled by the use of foam suppressors.

Spray cleaning machines or "metal washing machines" are often used to clean parts between processing steps when the part must be clean prior to further mechanical processing. The machine will clean, rinse, dry, and cool (if necessary) so that parts can be taken immediately to a following operation. Spray cleaning is less often used as a part of a plating cycle. When parts are to be plated they are, of course, racked and the racked parts are then easily cleaned by soak or electrolytic cleaning. This does not, however, rule out spray cleaning—it may easily be the best preclean or provide the means to solve a particular problem.

Ultrasonic Cleaning

Ultrasonic cleaning depends on cavitation action brought about by the impingement of high-frequency sound on the surface of the work. The ultrasonic energy is supplied by a transducer or a generating source that radiates the energy onto the surface to be cleaned. The cavitation that takes place on the surface of the work provides very effective agitation. The energy is released on the surface it sees, and it will see into very small crevices and holes.

Small and intricate shapes that are caked with dirt may be cleaned by this method when all others fail. Areas to be cleaned must see the radiating surface and should be placed close to it to take advantage of the radiation.

In general, the same solvents and solutions can be used for this type of cleaning as for other types. However, temperatures and concentrations will probably be much different. The cavitating action is limited by the vapor pressure of the cleaner, which in turn is limited by the temperature.

It is possible to clean large areas by this method by the use of a scanner. However, such a device is complex, expensive, and slow in terms of total area cleaned.

Selection of a Cleaner

The most popular method of selecting a cleaner consists of consulting a local sales representative. There are a number of advantages to this approach. The representative knows his products, knows cleaning problems, and has a product to satisfy every need. Also, he frequently points out a few facts that did not occur to the customer. In addition, there are a number of distinct advantages to the propietary product. It comes in a package, premixed to avoid storage of component chemicals. It is formulated for easy handling—free-flowing, non-caking, and dustless. Instructions are available that describe the cleaner and recommend typical applications and specify make-up, control, and operating procedures. A simple chemical control kit or a control procedure is made available.

There are a few things that favor cleaner formulation: The component chemicals are the least costly in the unmixed form and the cleaner composition will be known.

It is a distinct advantage to know what chemicals are present. There is a function for each chemical and it is helpful to know what these are and what they are expected to do. More important—control can be done in terms of the components or at least an approximation there of rather than on an empirical basis.

Whether cleaners be proprietary or formulated it is well to test a number of them. This will do two things. First, it will allow selection of the best cleaner. Second, it will establish a preliminary test history.

Control

A common method of cleaner control consists of maintenance to prescribed limits based on a simple titration. A used cleaner and a fresh cleaner at the same titration value will not contain equal quantities of alkali. This, however, is of little consequence since the cleaning time is set to assure cleaning with the aged cleaner. Thus, a simple control procedure and fixed cleaning conditions are allowed by using an excessive initial cleaning time. The loss in cleaning value can be established by a cleaning test. Such tests cannot be reproduced closely, but a test history and retesting can be very useful to establish cleaner life and to correlate the control method and maintenance procedures.

An occasional complete chemical analysis is helpful to understand the cleaner, the response to control, and how it decomposes with use. Total alkali, phosphate, silica, carbonate, and soap can be determined gravimetrically without too much difficulty. Complex phosphates, borates, surface active agents, and wetting agents are somewhat more difficult.

Routine control methods are empirical. A single end-point or a double end-point titration are usually made to estimate the effective alkalinity. Titration to an alkaline end-point, such as with phenolphthalein, is sometimes used as an approximate measure of the cleaner concentration. The empirical value is then established by titration of a fresh solution of cleaner of known concentration:

$$\frac{\text{cleaner concentration}}{\text{titration}} = \text{factor}$$

A second titration to an acid end-point, such as methyl orange, is often made and the cleaner is maintained to the two values. This is done by additions of strong alkali to maintain the alkaline end-point and additions of cleaner to maintain the acid end-point.

Titration methods should be selected to satisfy the cleaner: A method that applies to a light duty cleaner is likely to be completely inadequate for a heavy duty cleaner. A simple single titration may be unsatisfactory. Also, titration with common available indicators may not properly express the strong alkali and the total alkalinity.

Titration with the aid of a pH meter is revealing. A pH–titration curve of a fresh cleaner sample may show significant breaks in the curve that will establish pH control points. A cleaner that consists of sodium hydroxide, sodium phosphate, and sodium carbonate lends itself to reasonably accurate control by titration to definite pH end-points. The NaOH in this cleaner can be determined by titration to pH 11.3. Titration to pH 10.4 will measure NaOH and ⅓ of Na_3PO_4. Titration to pH 8.5 will measure NaOH, ⅓ Na_3PO_4, and ½ Na_2CO_3. Titration to pH 4.2 will measure NaOH, ⅔ Na_3PO_4, and Na_2CO_3.

When a pH meter is used, each of these end-points easily can be observed. From such an analysis a control equation can be derived and each chemical that is needed can be added. In a cleaner of this type the NaOH will decompose and must be replaced. Na_3PO_4 will not decompose as long as NaOH is present but will only be lost by drag-out. Na_2CO_3 should not be added since it is formed in the process. Proper control reveals that NaOH should be added frequently, Na_3PO_4 occasionally, Na_2CO_3 not at all. Addition of cleaner rather than the component chemicals would be uneconomical by comparison.

A pH curve on a silicated cleaner will not reveal breaks that separate the silicate from the NaOH and the Na_2CO_3. However, NaOH can be approximated from a highly alkaline end-point and silicate from a less alkaline end-point.

Cleaners can usually be controlled by the empirical assumption that the cleaner consists of free alkali and combined alkali. These two should essentially consist of NaOH as free alkali and sodium phosphate or sodium silicate as combined alkali. In mild cleaners the carbonate may be considered as combined alkali. Any method that will reasonably approximate these empirical definitions will provide control. The definitions, however, must have some correlation with cleaner performance and cleaner life. A properly chosen double titration will usually suffice.

When the cleaner is a medium duty or a light duty cleaner, "free" alkali may be defined as partly sodium phosphate (or silicate), and the combined alkali as the remainder of the phosphate, or silicate, and a part or possibly all of the carbonate. With use, such a cleaner will lose alkalinity, which may be replaced by addition of NaOH even though no NaOH was present initially. This should only be done, however, with a complete understanding of the decomposition and replacement chemistry. Otherwise, a medium duty cleaner might inadvertently be converted to a heavy duty cleaner.

A measurement of specific gravity can be of value if it is supplemented by a titration or other measurements. It essentially reveals the total salt present, indicating how the total concentration of the cleaner increases due to decomposition when the free and combined alkali are maintained.

Conductivity measurements have been found useful and are responsive to the strong alkalis that are important to maintain the cleaning power. Simple procedures for control have been established based on conductivity or conductivity supplemented by another measurement. Conductivity adds an additional control factor for electrolytic cleaners. When the conductivity is controlled the current density can be maintained at an original set value without a change in voltage. It is well also to note the converse: that the necessity to increase the voltage of a cleaner is an indication of loss in alkalinity.

Surface tension tests and foaming tests have been found helpful and occasionally even essential to complete control.

7. PICKLING

No two plating processes are ever the same. Those who are responsible always change them to some small degree. This may be a matter of experience, prejudice, or personality. Frequently, however,

Table 7 Pickling Solutions

Acid Concentration, oz/gal					*Metal Pickled*	*Temp., °F*
H_2SO_4	HCl	HNO_3	HF	CrO_3		
5–10					iron	70–170
20–35					copper	70–140
	5–10				iron	120–140
	10–20				iron	70–100
		60–80			silver	70–120
			5– 6		lead	70–100
			15–20		magnesium	70–100
5–10		2– 4			iron	70–100
3– 6		7–10			magnesium	70–100
10–20		4– 8			iron	70–100
30–45		10–15			copper	70–100
5–10	2– 4				iron	70–140
8–10			4– 5		st. steel	70–140
		8–10	1– 2		st. steel	70–140
		15–20	1– 2		titanium	70–100
		25–35	2– 4		nickel	140–170
		20–25	20–25		st. steel	120–150
		70–90	8–12		aluminum	70– 80
		10–14		12–16	magnesium	70–100
		2– 3		20–24	magnesium	70– 90
.05–.10				20–24	magnesium	70– 90
40–44				8–12	aluminum	100–180
2– 3				30–40	zinc	70–100
	3– 5			60–80	zinc	70–100
	5– 7	50–60			nickel	70–100
			1– 2	6– 8	st. steel	70–160
15–20	6– 8	8–12			nckel	160–180
60–70	0.2-0.4	50–60			nickel	120–160
25–35			4– 6	8–14	aluminum	140–150

The publisher acknowledges permission to reprint this chapter from the November 1968 issue of *Metal Finishing*, Westwood, N. J.

it is a matter of necessity to compensate for a difference in the structure, the roughness, or the surface oxidation of the metal. A pickle is one of the treatments that is often changed.

The pickle to use is the one that does the job; but the pickle to start with is the one that has been well established by experience.

A number of pickles are listed in Table 7.

Pickling Acids

In the absence of experience, the first acid that should be tried is sulfuric acid. This is cheap, non-fuming, stongly acidic, and forms soluble salts with most metals. Steel can be pickled with a concentration of 5 to 10 ounces per gallon. Removal of light surface stains, or deoxidizing, is done at room temperature. Activating and descaling is done at higher temperatures. At higher acid concentrations, copper is deoxidized.

Sulfuric acid will activate plain low carbon steels. Medium carbon steels form a carbide smut when pickled that is less severe with hydrochloric acid. At 5 to 10 oz/gal steel can be activated in a few minutes at 120 to 140°F. Room temperature pickling can be done at higher concentrations. Hydrochloric acid gives off disagreeable fumes, but these are less noticeable in dilute solutions, even at elevated temperatures. The acid forms salts with most metals that are more soluble than the sulfates.

Nitric acid, due to its oxidizing ability, will dissolve resistant and noble metals such as silver.

Hydrofluoric acid is useful to pickle lead at low concentrations and magnesium at higher concentrations.

These four acids, plus chromic acid, when used in various combinations, will satisfy the majority of pickling needs.

A great many of the pickles are made up with two acids. Sulfuric–nitric pickles of various ratios are used to treat medium-carbon, heat-treatable steels, magnesium, and copper. These pickles not only are more economic than nitric acid but they also produce a cleaner surface than the component acids.

Nitric–hydrochloric is a compromise pickle that has some of the advantages of each of the acids when used on steel.

Hydrochloric acid is not used in high concentrations in combination with other acids because of the high vapor pressure of hydrogen chloride and because it will decompose in the presence of nitric acid. Fortunately, hydrofluoric acid is more stable and less volatile and so is a good substitute for hydrochloric in a mixed acid.

Nitric–hydrofluoric acids will pickle stainless steel, titanium, nickel, and aluminum.

Hydrofluoric acid should be classified as a "nasty" acid. It burns the skin readily and then poisons the flesh. It is used when nothing else will work but should never be used when there is a good alternate.

Chromic acid has little popularity as a chemical pickle but is much used as a strong oxidizing acid when supported by other acids. A number of "chrome" pickles are beneficial to the treatment of magnesium—for example, chrome–nitrate and chrome–sulfate. Chrome–fluoride will pickle stainless steel.

It is evident that "mixed-pickles" made up of several acids offer many advantages. Mixtures of three acids have been found useful, as illustrated by the sulfuric–hydrochloric–nitric pickles for stainless steel and the sulfuric–hydrofluoric–chromic pickles for aluminum.

It is inconvenient to have to use a great number of pickles in a wet processing shop. Table 7 would indicate that there is little hope of doing many different pickling jobs with a few pickles. On the other hand it is obvious that many of the formulations are quite similar. One nitric–hydrofluoric pickle can be devised to process steel, titanium, nickel, and cobalt. Copper can also be treated in this same bath but excessive treatment will result in contamination with dissolved copper and resulting "copper-flash" on subsequently pickled less noble metals.

The five acids of Table 7 provide great versatility in pickling. However, many other acids and salts are used because of special chemical properties.

Phosphoric acid is a useful commercial mineral acid. This acid forms soluble salts and has also an ability to form films that "passivate" the work. Its properties make it useful as a component in bright dips and as an etchant suppressor.

When a pickle is selected, the metals present in the alloy and the solubility of the salts formed with the pickling acids should be considered. For example, if the pickle contains sulfuric acid and lead alloys are treated, then white films of lead sulfate will form on the work. Unfortunately, the effectiveness of a pickle cannot be predicted from solubility data. Problems are solved by trying known and recommended pickles and by experimentation. When the common acids do not produce results the less common ones should be tried.

Organic acids have occasional use—acetic acid has been found useful in bright dips and pickles for lead alloys. Tartaric and oxalic acid find limited use.

Fluoboric acid forms highly soluble salts with most metals and is therefore a good candidate as a general purpose acid.

Oxidizing salts are quite commonly used in mixed pickles. Sodium dichromate is an easily handled salt that is convenient to make up chrome–sulfate pickles. Ferric sulfate with sulfuric acid is used to pickle copper alloys, while cupric chloride and hydrochloric acid is used to pickle nickel alloys.

Neutral salts are used as a source of activating ions. Sodium chloride is added to other acids to form small amounts of hydrochloric acid. Sodium sulfate, sodium nitrate, and sodium fluoride are used in the same way.

Ammonium bifluoride serves as a source of acid fluoride when a mixed acid containing fluoride is needed. It is welcome because of a much greater ease of handling than hydrofluoric acid.

Hydrogen peroxide is used when a very strong oxidizing action is desired. It will promote pickling of weak acids and even weak alkaline solutions. It is beneficial in various bright dips and when mixed with organic and inorganic acids. Also it is used with cyanide and with ammonia to pickle copper alloys. It does however decompose quite readily and is an expensive ingredient.

Pickling

Pickling in a broad sense is the acid removal of metal from a surface. In a narrower sense, metal is removed from a surface for a number of reasons: Scale must be removed from a surface before the surface is plated. Scale can be removed with acid by pickling, which can be called "acid descaling". Scale that has been formed by hot rolling, annealing, or heat treating is economically removed by this treatment. The metal is usually pickled at the mill so that descaling is not normally a part of the preplate treatment. However, the metal may be purchased in the scaled condition.

When steel is pickled it is desired to remove scale with as little removal of metal as possible. Steel is pickled at the mill in inhibited acid. Inhibitors are compounds such as thiourea or sulfonated quinidines that greatly retard the attack of steel by hydrochloric or sulfuric acid. The direct action on the scale is not retarded, so the action on the metal is therefore much less than in uninhibited acid. Inhibited acids are essential to prevent excessive pitting of heavily scaled steel parts during pickling.

Inhibited acids have little use in plating shops for several reasons: Steels with heavy scale are seldom pickled. A pickle in a plating shop is generally used to activate by removal of metal, and inhibitors are therefore undesireable. Inhibitors can be dragged through the rinse and into the plating tank where they cause adhesion problems. The inhibited pickle should, however, be kept in mind as an economic means of heavy duty acid descaling when needed.

After reactive metals are pickled, an oxide re-forms on the surface. At room temperature, a thin "natural" oxide will form, while a light scale will form during mild heating due to annealing, forming, or machining of the metal. These thin or light oxides along with sulfide stains that form on copper and silver must be removed prior to plating. This is done by light pickling or acid dipping.

If a part is pickled, rinsed, and then introduced into a plating bath it usually will not form a film of sufficient thickness to interfere with plating adhesion. If the time, on the other hand, is long or the part is allowed to dry it may then require a light acid dip.

What has just been said leads to the point that metals may require various treatments—a strong pickle, a mild pickle, or an acid dip. This also is true because of other factors in addition to the amount of scale, oxide, or stain. Some alloys are much more resistant to acids than others. Cold rolled metals sometimes have embedded impurities that must be removed or they have a severely cold worked surface layer to which electroplates will not bond. Because of all these variations Table 7 is only a guide. In fact, in view of the differences in metals, and due to mechanical, thermal, and chemical history of the metal, any pickle should only be regarded as a recommendation.

Regard the pickle with suspicion. On the other hand, be assured that when the quality and condition of the metal is controlled and the pickle has been established, the practice is dependable. The great majority of working pickles have been used year in and year out with no change in the control limits or the working practice.

Pickling Terms

The following list of pickling terms will be useful in the identification of process steps:

Pickling (as a general term)—includes all acid and some alkaline processes that remove metal from the surface.

Pickling (as a specific term)—an acid process for the removal of scale.

Descaling (as a general term)—a process for removing scale by mechanical or chemical means. Descaling by the use of acid as we have seen is "acid descaling" or pickling.

Deoxidizing—a chemical treatment that removes the natural oxide, a light scale, or a stain from the surface. An acid that will remove oxide without attacking the underlying metal is a true deoxidizer.

Acid Dipping—a short dip in an acid that will remove light oxides, stains, or films that form in the rinse or due to excessive time between steps.

Neutralizing—an acid dip used to neutralize alkaline films that persist after alkaline cleaning and rinsing. This need only be done with a very dilute acid solution.

Desmutting—an acidic treatment that will reduce or remove the amount of smut on a surface that has formed in the previous chemical step and that is not removed by the rinse.

Activating—any treatment that is essential to chemically prepare the surface so that the subsequently applied deposit will bond to that surface. These treatments are usually acid treatments.

Acid Cleaning—an acidic pickling step that aids in the removal of contamination of unknown debris on the surface that must be removed in order to process the part.

Quite often one acid step is all that is needed. This step may descale and activate; it may be a simple activating dip, it may also desmut and clean. Pickles often do more than one job, but if it is the only acid step used then it is well to merely call it a "pickle."

Electrolytic Pickling

Metal is easily and reliably removed from a surface by making it an anode. As an example, steel can be electropickled in a dilute solution of sulfuric acid. The amount of metal removed will depend on the current density and the time. Some metal will be removed chemically, but in a dilute acid bath this will be small compared with the electrolytic action. The method is advantageous when it is desired to remove a precise amount of metal or to remove metal in a fixed time.

Anodic pickling will descale, pickle, and activate. However, it can also etch, promote smut, and passivate. Just as in chemical pickling, there are many solutions that can be used and a wide variety of reactions can result—depending on the composition of the bath and the alloy being pickled.

A metal will generally dissolve anodically in an acid that will form a soluble salt of that metal. The reaction is readily controlled by the current. Only enough acid is needed to keep the work from polarizing, just as in a plating bath. A part can in fact be electropickled in a plating bath, although the effect of the dissolved metal on the bath may be deleterious to further plating.

If metal is to be removed anodically it can of course be removed in solutions other than acids, particularly cyanides.

Electropickling is done with alternating current, periodic reverse current, and cathodic current. In an acid solution of sufficient strength the metal will pickle chemically and the electrolytic action will be in addition to the chemical action. As an anode the action will be chemical plus electrochemical. As a cathode the part will still pickle chemically and the cathodic current will not inhibit this action. In many solutions, of course, metal will plate at the cathode.

Cathodic pickling generally has little effect on the work. It can, however, desmut, reduce etching, and sometimes activate. It is interesting to note that the work can be activated both as an anode and as a cathode.

Desireable and undesirable actions can take place at either electrode. When an undesirable reaction takes place at an electrode, a corrective action might take place at the electrode of opposite charge. For this reason, alternate anodic and cathodic pickling has solved some problems. Alternating pickling, in some instances, has overcome problems with smut, etching, or polarization. The application of current to a pickling bath greatly increases the possibilities of changing pickling charateristics.

Electropolishing

Most metals can be electrochemically polished by anodic treatment in concentrated acid solutions. This process is closely related to electropickling but the polishing action can only be maintained under much more closely controlled conditions than those allowed during electroetching. The solution, the temperature, the current, and the time all must be controlled within narrower limits.

When it is desirable that a part be smooth in addition to being pickled, then an electropolishing step can be advantageous. Generally, the process is used to smooth or brighten the work and pickling aspects are of little importance due to the fact that these methods are quite expensive compared with chemical or electrochemical pickling.

Chemical Polishing

Many metals can be polished or at least brightened by acid treatments. These treatments are done in dilute and concentrated acids, the formulations being more strict than the etching pickles. Chemical polishing, chemical brightening, and bright dipping are all used to produce smooth or bright surfaces by acid treatment.

Chemical Milling

Quite a number of processes have been developed to chemically reduce the dimensions of metal in competition with mechanical processes. Chemical milling, chemical maching, chemical blanking, and chemical deburring are processes that have reduced costs where special problems exsist, such as removal of metal from complicated shapes, areas that are difficult to reach with a tool, or metals that are hard to machine. Similar electrochemical processes have also proved fruitful. These solutions are formulated to avoid etching. Interestingly enough, they are also formulated to avoid excessive polarization since sizing is more faithful in the absence of polarization. It is apparent that these solutions lie somewhere between the pickles and the polishing solutions.

8. STRIKE PLATING

Strike plating or "striking" is a preparatory step essential to bonding or covering by a subseqently applied deposit. It is a means of plating the substrate and keeping it active at the same time. A strike step is vital when steel is plated with silver. Likewise, high efficiency copper cyanide can only be bonded to steel after it is struck.

Cast iron cannot be directly plated with zinc from a cyanide bath. This problem is solved by striking the cast iron with copper. Here the function of the strike is different. It merely provides a surface on which zinc will deposit rather than just hydrogen.

Low efficiency alkaline and cyanide baths promote covering of the substrate. Subsequent plating sometimes will cover better over one of these strikes.

Covering and adhesion of plating is promoted on zinc and on zincated aluminum through an intermediate copper strike.

The need of a strike step is established by experimentation. Consider a basic plating cycle that consists of (1) clean, (2) rinse, (3) etch, (4) rinse, (5) plate, (6) rinse. There are many deviations from this cycle. When deposits are thin, steps 3 and 4 may be eliminated. If deposits are heavy and bond is important, a precleaning step might be needed. In plating cycles such as plating on aluminum the total number of steps doubles and the process becomes elaborate.

Another deviation from the simple cycle is the insertion of a strike step between steps 4 and 5.

A strike sometimes will help overcome a deficiency in material quality. Partially decomposed oil worked into a surface can cause spotty, non-adherent, or pitted plating. Although it is not the best answer, a strike may solve this problem.

There is no fixed formula for a strike bath. A low concentration bath may be used with a long striking time or a higher concentration for a shorter time. The cathode efficiency should be low, however, so that profuse gassing takes place during deposition.

Sometimes more than one strike is used—a dilute strike followed by one more concentrated. This is quite common prior to silver plating of steel.

Copper strikes are widely applied. The copper strikes and the low efficiency copper cyanide baths are one and the same: cyanide copper plating as a pre-plate functions as a strike.

By increasing the temperature and adding trisodium phosphate a copper strike can serve as a strike and as an electrocleaner. It is not recommended as a substitute for cleaning but it adds cleaning insurance and may eliminate a precleaning step.

Low efficiency alkaline tin baths have the characteristics of a strike and are so used, although not commonly.

Acid nickel baths are sometimes called strikes and are used in a similar manner. It has been found, for instance, that silver plating of steel was more consistently bonded when the steel was plated with nickel then struck with silver prior to silver plating. The nickel strike, however, cannot be used as a substitute for a silver strike and so does not promote bond but merely increases reliability in some processes. Nickel does act like a strike prior to plating on stainless steel. Properly formulated nickel baths will keep stainless steel active while covering it with a nickel deposit. This process shows us that a strike may be regarded as a simultaneous plating and activating bath. It is the step that keeps the metal receptive to bonding while depositing another metal that is less sensitive to passivation.

(See also p.29 regarding striking for activation.)

9. RINSING

To rinse properly, all that is required is to use a sufficient excess of water. This satisfies the need to remove solution from the surface of the work and to replace it with clean water before proceeding with further processing in the plating line. Water is cheap and plentiful so the problem is not serious. As a result it is not surprising that rinsing is frequently an uncontrolled plating line variable.

The Running Rinse

Common rinsing practice is to provide rinse tanks and to allow the water to flow. It is now realized that this literally amounts to dollars down the drain. The safe excess of water becomes costly when great amounts are used merely to be sure that the water is clean, and unfortunately water is not always available in large quantities.

Actually we do not rinse with clean water but with water with an acceptable level of contamination.

At the highest allowable level of contamination, soluble chemicals are most efficiently removed from the work, because as we shall see the rate of removal of soluble chemicals is directly proportional to the concentration in the rinse. At half the concentration it will take twice as much water to remove the same total quantity of chemicals. If a rinse tank is allowed to run without processing work the concentration will drop markedly. The rinsing efficiency will likewise decrease. The economy of rinsing is well worth our concern when it is done on a sufficiently large scale.

Purpose

The purpose of rinsing as applied in electroplating lines is to reduce the soluble chemicals to a contamination level acceptable to the next tank or the next step in the process.

Methods of Rinsing

The common rinse is a single tank with a flow of water through it. This is a simple immersion rinse that may be followed by another stage or supplemented with other types of rinsing. An important requirement of such a tank is to introduce the water at the bottom. In very large tanks the water should be introduced at a number of points and allowed to overflow along the full length of the tank. Multiple rinse tanks in two or more stages or counterflow rinse tanks will cut the water consumption substantially. A spray rinse is recommended if a single rinsing station is to be used to rinse away different types of solutions. Spray and fog rinses are used when it is desired to rinse with small quantities of water for a short period of time.

Rinse Control

Substantial savings can be reälised by the use of multiple rinse tanks as well as by control of water flow. The rinse easily can be controlled by throttling the water inlet valve. It is better controlled with a flow meter and a valve and best controlled automatically. Instruments are available to do this.

A commercial rinse tank controller consists of a Wheatstone bridge with a cathode ray tube in one leg of the circuit to indicate balance. An off-balance in either direction will activate or deactivate a relay to affect warning or controlling devices. In a typical installation the instrument will open a solenoid valve when a set concentration is exceeded and close the valve after enough water has flowed to drop the concentration to the pre-set value just below the balance point. The controller was developed principally to save water costs; one large manufacturer reported savings up to 97%. A laboratory reported a controlled rinse tank running for 8 minutes that would otherwise have run for eight hours as an uncontrolled rinse. The instrumental method was applied purely for economic reasons. After it was available and working it then became practical to consider the advantages of the controlled rinse.

Rinse tank controllers have been applied to automatic equipment where large quantities of water would be costly if uncontrolled. This is in keeping with sound engineering practices that must be applied to large scale processing.

Rinsing Practice

Typical plating sequences consist of: clean, rinse, pickle, rinse, plate, and rinse. After the cleaner the work is rinsed to avoid neutralization of the pickle. After the pickle the work is rinsed to avoid contamination of the plating tank. And after plating the work is rinsed to avoid spotting or staining. Possibly one rinse could be used in place of all three if it were kept sufficiently clean. However, in economic practice, alkaline rinses are kept seperate from acid rinses and the final rinses are kept cleaner than the intermediate rinses.

When rinse controllers are available it becomes necessary to decide on a contamination level for the rinse in order to set the controller. The level of contamination varies with the local water as well as the type of work being plated. General figures have been established that set a controlled concentration of 0.1 oz/gal on rinses following cleaners and pickles and 0.02 oz/gal on final rinses. These values however should only be regarded as starting figures. A working value is established by gradually increasing the controlled concentration and noting the effect on the work or on the following tank.

Rinsing Efficiency

Since rinsing efficiency is directly dependent on the concentration it is desirable to operate the rinse at as high a concentration as possible without decreasing the acceptable quality of the work. A study of drag-in is profitable to appreciate and resolve this problem—a study that is readily made with the aid of a conductivity meter. Such studies are important because the rinse concentration is only half of the story. The other half is the drag-out from the previous tank that depends greatly on the drain time and sometimes on the method of racking. On automatic equipment such studies have revealed that mechanical shaking of the part when in the drain position should be employed. A by-product of these studies is a savings in plating solution by reducing drag-out.

Rinse Calculations

The concentration of salts in a rinse tank will reach an equilibrium at steady processing conditions. The equilibrium will depend on a fixed flow rate, regular introduction of work, and sufficient time to reach equilibrium. The concentration will be equal to the salt introduced per interval of time divided by the water introduced in the same time interval:

$$C_r = \frac{S_r}{W_r} \tag{1}$$

where C_r is the concentration in rinse, S_r is the salt into rinse per time interval, and W_r is the water into rinse per time interval, and

$$S_r = V_d \times C_s \tag{2}$$

$$W_r = F + V_d \tag{3}$$

where V_d is the volume of drag-in per time interval, C_s is the concentration of salt in drag-in, and F is the water flow.

In (3), V_d is small compared with F and may be ignored. From (2) and (3), (1) becomes

$$C_r = \frac{V_d \times C_s}{F} \tag{4}$$

We now introduce a term R and rewrite (4):

$$R = \frac{F}{V_d} = \frac{C_s}{C_r} \tag{5}$$

where R is the "rinsing ratio."

From (5), three equations follow that are very useful:

$$C_r = \frac{C_s}{R} \tag{6}$$

$$F = R \times V_d \tag{7}$$

$$F = \frac{C_s \times V_d}{C_r} \tag{8}$$

Example 1

Work is being introduced into a tank at a rate of 10 sq ft/min. The drag-in has been determined to be 0.2 fl oz/sq ft and the concentration of salts in the drag-in solution is 8 oz/gal. What flow rate is required to maintain an equilibrium concentration of 0.1 oz/gal in the rinse? Using (8):

$$F = \frac{C_s \times V_d}{C_r}$$

$$= \frac{8 \text{ oz/gal}}{0.1 \text{ oz/gal}} \times \frac{0.2 \text{ fl oz/sq ft}}{128 \text{ fl oz/gal}} \times 10 \text{ sq ft/min} \times 60 \text{ min/hr}$$

$$= 75 \text{ gal/hr}$$

The same problem solved for the rinse concentration is as follows.

Example 2

Work is introduced into a tank at 10 sq ft/min and drag-in is 0.2 fl oz/sq ft. What equilibrium concentration will result at a flow of 75 gal/hr? Using (4):

$$C_r = \frac{V_d \times C_s}{F}$$

$$= \frac{0.2}{128} \times 10 \times 60 \times \frac{8}{75}$$

$$= 0.1 \text{ oz/gal}$$

Counterflow Rinsing

Large quantities of water can be saved by countercurrent rinsing. In this system the water flows in a direction opposite to the direction of the work through a series of two or more tanks. In a two tank system the work is rinsed first in No. 1 and then in No. 2. The overflow from No. 2 flows directly into No. 1. The No. 2 tank will be the least contaminated. The rinse equations for counterflow rinsing become:

$$F = R^{\frac{1}{n}} \times V_d \qquad (9)$$

$$C_r = \frac{C_s}{R^n} \qquad (10)$$

where n is the number of counterflow tanks.

When the above problems are applied to a two stage counterflow rinse the answers are as follows.

Example 1 (double rinse)

$$R = \frac{C_s}{C_r} = \frac{8}{0.1} = 80$$

$$F = R^{\frac{1}{n}} \times V_d$$

$$= 80^{\frac{1}{2}} \times \frac{0.2}{128} \times 10 \times 60$$

$$= 8.4 \text{ gal/hr}$$

Example 2 (double rinse)

$$R = \frac{F}{V_d} = \frac{75}{\frac{0.2}{128} \times 10 \times 60} = 80$$

$$C_r = \frac{C_s}{R^n} = \frac{8}{80^2}$$

$$= 0.0012 \text{ oz/gäl}$$

A number of interesting points are brought out by these calculations:

Example 1 applied to the double rinse shows that water consumption can be cut to 11% of the single rinse requirement.

Example 2 shows that at the same flow the concentration will become 1.2% of the concentration in a single rinse when using the same amount of water.

When the rinsing ratio R is used it becomes easy to estimate the effect of another rinsing stage merely by increasing the value of the exponent n.

Rinsing Evaluation

The following steps are recommended as an approach to the rinsing problem:

1. Decide on the desired limiting concentration in the rinse.
2. Measure drag-in by conductivity measurement, and from the

drag-in per rack estimate the required flow from the rinse equation.

3. If the flow is appreciable, estimate the economies of a counterflow rinse.

4. Install a flow meter so that a definite flow can be maintained.

5. Calibrate the rinse tank with a conductivity meter so that concentration in the rinse can be measured.

6. After the flow meter and the rinse tank are calibrated, determine the characteristics of the rinse by measuring the change in concentration with time, under full production conditions, by means of conductivity measurements.

7. After the rinse has been in service for some time, study the plating log to determine if the rinse tank is a source of contamination to the work or to the next tank in line.

8. Calculate the savings when operation is kept at equilibrium and estimate the need for continuous control with automatic equipment.

Rinsing Problems

Close packed parts, as in barrel plating, take some time to rinse. Experiments with barrel rinsing have readily shown what goes on. If a barrel is allowed to rotate in a running rinse while conductivity readings are taken, it will be seen that conductivity will increase for some time. During a typical study this time was two minutes, revealing the time necessary to rinse the barrel. With automatic equipment the controller can signal when rinsing is complete or the barrel and the rinse flow can be halted. This assures proper rinsing in the minimum amount of time. Also, it was found with this approach that rinse flow rates could be increased to shorten the rinsing time.

With automatic control a combined process step is possible that consists of a holding rinse. The controlled rinse will allow water to flow until the equilibrium concentration is established and then shut off the water to maintain this concentration. Following cleaning, some steels can be held in the inactive alkaline rinse to protect the metal until further processing is required.

After pickling, metals may be held in an acid rinse to maintain an active surface and to avoid rusting and so that adherent plating will be attained.

It is believed that troubles can develop from rinse water that is too clean. Rinse waters usually consist of calcium bicarbonate plus contaminants. The calcium bicarbonate can act as a neutraliser both for acids and alkalis and the rinse can become alternately acid and alkaline if it is allowed to run to sufficient dilution. It is possible that metals can lose activity under such conditions or, conversely, that the metals may corrode.

10. ANODIZING

Useful coatings are produced on some metals by anodic oxidation —by treating the metal as an anode in a selected electrolyte. Commercial coatings are produced—on a large scale on aluminum and magnesium and on a much lesser scale on zinc, titanium, and some of the other minor metals.

Aluminum

A remarkable nonmetallic coating is produced by anodizing aluminum. The coating is hard, corrosion resistant, and nonductive. The commonest coatings are porous and can be dyed any color. These properties offer a welcome alternate to the electrolytically produced metal coatings that adds dimension for decorative as well as engineering adaptations.

Durable anodic coatings are applied to architectural hardware and building panels to preserve the natural finish or to add a lasting tone, produced by variations of the process. Colored finishes are used on cameras, sports equipment, appliance trim, electronic connectors, and novelties. Anodic films provide insulation. A film 0.5 mil thick on 1100 alloy will have a break-down voltage of 500. Much higher insulation voltages can be attained on high purity alloys or by the application of "hard" coatings. Thin glass-like transparent coatings can be applied to protect aluminum reflectors.

Chromic Acid Anodizing

Chromic acid anodizing by the Bengough and Stuart process[1] was applied for the protection of seaplane parts. The process uses a 3% solution of chromic acid at 100 to 115°F. The work is placed in the bath with no current applied and the voltage is raised to 40 volts during the first 15 minutes; it is then raised to 50 volts and the work anodized for a total time of one hour. The current density is 2.5 to 4 amp/sq.ft. Military Specification MIL-A-8625[2] requires a minimum current density of 1 amp/sq ft that can be maintained with chromic acid up to 10% or greater with additions of H_2SO_4 up to 80 g/l.

Chromic acid coatings are relatively soft as compared to sulfuric acid anodizing. They are usually a light gray on the alloys that are relatively pure and a darker gray with higher alloying elements present. Most processing is very similar to the original process, producing coatings of 0.1 to 0.2 mil with good corrosion resistance. The coatings can be dyed and the corrosion resistance increased by sealing in hot water.

Sulfuric Acid Anodizing

The most used process is one that is operated with sulfuric acid. Many variations of the process are possible and are practiced, but commonly used conditions are similar to those specified in MIL-A-8625:

Coating	*% H_2SO_4*	*amp sq ft*	*°F*	*Coating wt., mg/sq ft*
non-dyed	15	12	68–72	600
dyed	15	16	68–88	2,500

Coatings of 0.1 to 1.0 mils that are clearer and harder are produced with the sulfuric acid process. Variations of this process, known as *Alumilite*, were developed by the Aluminum Company of America.

The 15% bath operated at 12 amp/sq ft and 70°F is sufficiently popular that it may be regarded as the "conventional" bath. The quality of coatings produced with this bath will be consistent when the bath is well controlled. The temperature of the bath should be held at 68 to 72°F by cooling (or heating if necessary). The current density should not vary from 12 by more than ± 1, which will not be difficult if the temperature is held. The bath voltage will be about 15.

The bath can be controlled by voltage when the exact relationship of voltage to current density is established. This is easily done by anodizing a simple shape at a current density of 12 and noting the voltage. A complex shape can then be anodized at this voltage with the assurance that the current density is right. The coating is dissolved in part as anodizing proceeds, so that aluminum and small concentrations of other impurities build up in the bath. The following limits are recommended:

	g/l	*oz/gal*
H_2SO_4	150–165	20–22
Al	0–15	0–2

Coatings of the same quality as those produced in the conventional bath can be formed at greatly different conditions of current, temperature, and acid concentration by the principle of compensating variables. Increased temperature, for example, increases the rate of dissolution of the aluminum, while a decrease in acidity tends to decrease this rate and an increase in current density increases the rate of formation of the film, thus tending to offset the more rapid dissolution rate at a higher temperature. When the variables are not compensating, but rather in the other direction, then a substantial difference in the hardness or softness of the film results. The character of the coating can be substantially changed by changing conditions.

The purity of the aluminum greatly determines the density of a coating from conventional baths. Dense, harder coatings are produced on high purity or commercially pure aluminum, while less dense and softer coatings are produced on the heat-treatable alloys that contain appreciable quantities of alloying elements. Copper as an alloying element is associated with lower density much more than silicon and magnesium. The anodizing characteristics of the commerical aluminum alloys have been well documented by the producers of the alloys.

Oxalic Acid Anodizing

Good quality abrasion resistant films can be produced in 5 to 10% oxalic acid at temperatures and currents similar to the conventional bath. However voltages of 50 to 65 are required and the films have a yellow cast. Substantially harder coatings are also quite easily produced in oxalic acid baths.

Hard Anodizing

Harder and thicker coatings are produced at lower temperatures and higher current densities. The MHC process of the Martin Co.[3] is a well known example. Coatings up to 6 mils in thickness are produced in 15% sulfuric acid at 32°F and 20 to 25 amp/sq ft. Voltages are started at 25 to 30 and increased to 40 to 60.

The Hardas process[4] utilizes a.c. current superimposed on d.c. at 60 volts in 6% oxalic acid.

Hard anodic coatings provide an extremely wear-resistant surface that exceeds the hardness of engineering chromium. Whether or not a coating is hard is, however, a matter of definition. One specification, AMS 2468A, specifies at a coating thickness of 0.002 ± 0.0005 inches a coating weight of 0.030 g/sq in, corresponding to a minimum density of 1.83 g/cc.

The density of a coating is relatively easy to determine by a strip-and-weigh procedure and correlates well with abrasion resistance. The coating can be assured to be hard or wear resistant by specification of the density. A minimum density of 2.2 or 2.4 is practical and commercial. Density control overcomes the problem of the variation of coatings properties as the alloying elements are changed. These variations are associated with the fixed anodizing conditions. With specified properties, particularly density, the baths can be changed to produce coatings that are more similar on different alloys.

Duplex Acid Baths

When two acids are used, some of the characteristics of each can be attained. Mason and Fowle[6] demonstrated that small additions of oxalic acid reduce the partial dissolution of the anodic coating by a sulfuric acid bath. This increases the hardness of the coating and extends the usefulness of the bath in several beneficial ways. A hard bath that can be operated at 50°F is made up with 12% H_2SO_4 and 2% $H_2C_2O_4 \cdot 2H_2O$.

At higher temperatures the sulfuric-oxalic baths produce coatings similar to the conventional coatings but with a slight yellow cast. The baths are more costly to operate than sulfuric acid because of the decomposition of the higher priced and less stable oxalic acid. They become economic when this cost is offset by savings in processing

Hard Sulfuric-Oxalic Acid Bath

	oz/gal	*g/l*
H_2SO_4	16–19	120–140
$H_2C_2O_4 \cdot 2H_2O$	1.3–2.7	10–20
Al	0–2	0–1.5
Temp., °F	48–52	
Current density, asf	30–40	
Volts	10–50	

time and cooling capacity. The cooling cost is appreciable in the low temperature baths and is a factor in the room temperature bath during summer months. Some anodizers add oxalic acid to a conventional sulfuric bath to allow operation at the same conditions, except at 85 rather than 70°F. This allows cooling during the summer, with tap water in place of refrigerant, to satisfy the dissipation of heat that is essential to proper operation. In the winter months the bath can be returned to conventional anodizing. This flexibility increases the dimensions of sulfuric acid anodizing.

Architectural Anodizing

The metal suppliers have developed several proprietary anodizing baths based on organic acids. These baths produce hard, light-fast, gray, black, and brown coatings that are durable and attractive, satisfying the exacting demands of large scale architectural applications.

REFERENCES

1. G. D. Bengough and J. M. Stuart, British Patent 223,994 (1924); U.S. Patent 1,771,910 (1930).
2. Military Specification MIL-A-8625, *Anodic Coatings for Aluminum and Aluminum Alloys.*
3. Glenn L. Martin Co., British Patent 701,390.
4. British Anodizing Ltd., British Patent 525,734.
5. Aerospace Material Specification AMS 2468A, *Hard Coating Treatment of Aluminum Alloys.*
6. R. B. Mason and P. E. Fowle, *Trans. Electrochem. Soc.* **101,** 53 (1954).

11. BRASS PLATING

Brass is a general term with which some liberties have been taken both by the metallurgist and the plater.

Brass alloys are red or yellow alloys that are made of copper and zinc and sometimes other elements in addition. When occasion demands, alloys are plated to match the color of the cast alloys. Electroplated brasses generally are composed of copper and zinc in proportions to give a brass color. White alloys that are principally zinc are also deposited and are known as "white" brass. Further, it is not difficult to co-deposit tin in addition to zinc, with the result that alloys can be formed that are as much a bronze (copper-tin) alloy as a brass. Generally, copper-tin alloys are considered to be bronzes, but a plated bronze is produced that consists of 8% zinc and 92% copper. This alloy is called a bronze because it resembles the bronze color of some of the cast bronzes.

Typical brass plating of a "brassy" colored alloy will be done with 20 to 30% zinc, the remainder copper.

Characteristics of the Brass Bath

Brass baths can be made by mixing a copper cyanide bath and a zinc cyanide bath. This is not a practical procedure but the fact that it can be done illustrates that brass plating consists of simultaneous deposition of two metals from a bath that has the combined characteristics of both of these baths. This knowledge, however, is only of limited use since these baths contain cyanide complexes that cause them to behave in a rather complicated fashion.

The alloys that are deposited from brass baths are true alloys, consisting of the same phases that are obtained by thermal means.

As an alloy process, brass plating is more limited than single metal plating. Specifically, the bath requires closer attention to control of the bath variables and more frequent checking of the bath composition (or checking of color reproducibility). (See Chapter 34 on Electroplated Alloys.)

Brass Plating

Brass is deposited to obtain adhesion of rubber to steel and for ornamental purposes. Various bath formulas have been proposed, but all commercial formulas are cyanide baths.

Formulation

The following bath has been recommended by Coats.[1]

	g/l	*oz/gal*
Copper cyanide	26.2	3.5
Zinc cyanide	11.3	1.5
Total sodium cyanide	45.0	6.0
Temperature	80–95°F.	
Current density	9 amp/sq ft	
Brass Anodes	25% Zn	
pH (colorimetric)	10.3–11.0	

The copper cyanide is a source of copper and the zinc cyanide a source of zinc. The sodium cyanide combines to form metal complexes with the following (assumed) formulations:

$$2NaCN + CuCN = Na_2Cu(CN)_3$$

$$2NaCN + Zn(CN)_2 = Na_2Zn(CN)_4$$

All sodium cyanide, in excess of this combining amount, is defined as free sodium cyanide. However, this is only by definition since the free cyanide cannot be determined by analysis. The excess cyanide above that required to form complexes aids in anode corrosion and increases the bath conductivity.

Carbonate will build up in the bath with use and will also increase the bath conductivity.

The anode and cathode efficiencies of the bath are both above 75%.

Bath Preparation

Steel tanks can be used satisfactorily for bath preparation and operation. Bath preparation consists of first dissolving the alkali cyanide and then adding the metal cyanides. The metal cyanides are soluble in and react with the sodium cyanide to form the complexes described above.

Operation and Control

The bath must be operated to maintain the desired composition in the deposit. For decorative purposes, this may be done by color control as follows:

A series of brass alloys are prepared covering a range of analyses, or a single brass plate of the desired color is used as a standard. The deposit is then compared with this color standard and if the colors do not match, the bath is corrected to obtain the proper color in the deposit.

For engineering purposes, the deposit is controlled by analysis.

Whether the deposit is controlled by analysis or color, a knowledge of the effect of the bath variables is essential.

If the deposit is high in copper, the percentage of copper may be decreased by:

1. Lowering the bath temperature.
2. Adding caustic to raise the pH.
3. Adding zinc cyanide.
4. Adding both zinc cyanide and copper cyanide to lower the free cyanide content.

If the deposit is high in zinc, the percentage of zinc may be decreased by:

1. Raising the bath temperature.
2. Adding sodium bicarbonate to lower the pH.
3. Adding copper cyanide.
4. Adding alkali cyanide to increase the free cyanide content.

It is best to hold as many variables as possible within defined limits, but it is not possible to define all limits on an alloy bath. Since the limits on the deposit are defined, as many of the bath limits as possible should be set with at least one variable to obtain the proper analysis of the deposit. This is the main difference in operation of an alloy bath and a single-metal bath.

In the single-metal bath, all limits can be defined. In cyanide baths, the cathode efficiency may then be a variable that will fluctuate depending on the condition of the bath at any given time. In the alloy bath, this condition cannot be tolerated. For instance, a drop in cathode efficiency in a brass bath may be due to a drop in the plating rate of the zinc alone. Consequently, the percentage of copper in the deposit will increase. Conversely, the plating rate of the copper may decrease and the percentage of zinc will increase in the deposit.

Another way of stating the problem follows: The analysis of the deposit cannot be accurately predicted from an analysis of the bath. The cyanide bath changes continuously; both carbonate and ammonia continuously form in the bath due to cyanide decomposition. The ammonia is driven off in the alkaline bath, but it is not driven off at the same rate as it is formed.

The bath should be operated in such a manner that great changes or large additions do not have to be made to correct the composition of the deposit. This can be accomplished by making frequent checks, by making small corrective changes, and by defining the limits of total metal concentration. If corrections result in an increase of both copper and zinc, then the bath should be diluted, but great dilutions should be avoided. The problem is really one of bath balance, as it is for single-metal baths, except that a balance has to be maintained for total metal and for the ratio of the metals in the deposit.

Temporary corrections can be made by changing the bath temperature or the current density. Increasing the current density will decrease the percentage of zinc in the deposit. This marks zinc as the noble metal in the recommended bath. However, in other brass baths[2] an increase in current density has resulted in an increase in the zinc in the deposit. This means that the nobility of the copper and the zinc can be reversed, so that for a given bath a log should be carefully kept and studied to determine the characteristics of the particular bath.

Plating-range tests will aid in control of the bath. If the deposit is off-color at a high current density, corrections may be made to obtain the proper color over a wide plating range. The deposit may turn red in the off-color range, a condition that may be due to either high copper or high zinc content. The metal responsible for the off-color deposit may be determined either by analysis of the deposit or by making an addition to the bath that favors deposition of one of the metals. If the test corrects the color, the cause has been found. If the test does not correct the color, additions favoring deposition of the other metal should be tried. After the troubles have been located and corrected in the plating tests, the corrections are applied to the production bath.

Changes in the Bath

The cathode efficiency may be increased by increasing the bath temperature or by decreasing the free cyanide. These changes will also affect both the ratio of metals in the deposit and the anode corrosion and should only be made in such a way that bath balance can be maintained.

Characteristics of the Bath

The cyanide brass bath has high covering and throwing power, but the deposit may go off color at extreme current densities. Therefore, closer attention to current distribution is required than for the single-metal cyanide baths if a uniform color is desired.

White Brass

A white brass alloy consisting of 28% copper and 72% zinc can be deposited from the following bath:

	g/l	*oz/gal*
Copper cyanide	16.8	2.2
Zinc cyanide	60.0	8.0
Total sodium cyanide	60.0	8.0
Sodium hydroxide	60.0	8.0
Sodium sulfide	0.4	0.05
Temperature	70–175°F	
Current density	10–100 asf	
Anodes	28% Cu	

White brass deposits are hard and corrosion resistant and may be used as a low-cost substitute for nickel when the degree of protection afforded by nickel is not required. The deposits have been used as an undercoating for chromium and as a white lacquer-protected coating on toys and metal trim.

REFERENCES

1. H. P. Coats, *Trans. Electrochem. Soc.* **73,** 435 (1938).
2. A. L. Ferguson and E. G. Sturdevant, *Trans. Electrochem. Soc.* **38,** 167 (1920).
3. F. Oplinger, *Proc. Am. Electroplaters' Soc.* **26,** 137 (1938).

12. BRONZE PLATING

Copper-zinc alloys containing 5 to 15% zinc are deposited to match the color of cast or wrought bronze alloys. In the range of 5 to 10% zinc the deposited alloys will be "red bronze," whereas higher percentages of zinc will give a lighter "bronze" color.

A. K. Graham[1] has defined the range and operating variables of a copper-zinc bath used for color plating that is partly described as follows:

Copper	oz/gal	4–6
Copper-zinc ratio in bath		
red bronze		95/5 to 90/10
light bronze		80/20 to 70/30
Zinc	as required	
Free NaCN	oz/gal	0.5–1.5
Na_2CO_3	oz/gal	2–8
Rochelle Salt	oz/gal	4–10
Ammonia (for color)		as required
pH		10.3–10.7
Temperature, °F		100–140
Current Density, asf		5–30
Agitation		mild
Anode (Cu-Zn)		90–10

Nathaniel Hall[2] describes a copper-cadmium solution that is recommended for deposition of bronze plating to attain more uniform color and minimize anode polarization:

	oz/gal
Copper Cyanide	3.0
Cadmium oxide	⅛–¼
Sodium cyanide	4.5
Sodium carbonate	2.0
Free NaCN	1.0
Temperature—room	
Anodes—copper	

Metallurgically, the copper alloys that contain tin are defined as bronzes. Plated alloys of similar composition are also considered to be bronze alloys.

Bronze, similar to brass, can be deposited from a mixture of copper cyanide and sodium stannate baths over the entire alloy range. The relative plating rates of the two metals cannot accurately be forecast but the influence of bath variables can be predicted from a knowledge of the component baths. Tin will not deposit at low temperature; it will deposit more readily as temperature is increased and will be suppressed by an increase in caustic content. The copper plating rate can be suppressed by an increase in the cyanide concentration and increased by an increase in temperature.

A bronze bath was developed by Battelle Memorial Institute[3] that will deposit a range of alloys. By variation of the tin in the bath, three distinct baths can be made that produce useful alloys as follows:

Approximate Composition			
Alloy	*Cu*	*Sn*	*Use*
1	93	7	Bearing surface
2	80	20	Undercoating
3	60	40	Decorative

Additions of tin to copper have a progressively hardening effect. Alloy 1 is used as a hard bearing surface. Alloy 2 is a harder but buffable alloy that is useful as an undercoating for other deposits. A white bronze that is very hard is produced by deposition of 40% tin in Alloy 3.

Bronze Baths

Three bronze baths that will deposit useful alloys and that illustrate the effects of bath variables by comparison are given here.

	1	*2*	*3*
CuCN, g/l	140	75	11
$Na_2SnO_3 \cdot 3H_2O$, g/l	45	37	95
Free NaCN, g/l	22	90	16
Free NaOH, g/l	8	8	15
Temperature, °F	160	170	150
Current density, asf	20	20	20
Anodes	bronze	bronze	two circuits
Tin in deposit, %	5–10	10–15	40–60

Bath 2 will deposit 1½ to 2 times the tin that Bath 1 will deposit with the normal variations allowed for a controlled bath. It takes a substantial difference in the baths to bring this about. To increase the tin in Bath 2 over Bath 1 takes a substantial reduction of the copper concentration as well as an increase in the free cyanide, both of which suppress the copper plating rate. In addition the increase in temperature favors the deposition of tin. This comparison points out two fortunate circumstances—that the bath can be changed to produce a second bath with a distinctly different tin content; and that on the other hand small changes would not greatly change the bath, so that control of the amount of tin deposited it practical within realistic ranges.

In Bath 3 the copper plating rate is suppressed in competition with the tin principally by a low copper concentration. Control of an alloy bath becomes more critical when the major metal deposited is present in a small concentration. A relatively small change in copper or in free cyanide can result in an appreciable change in the composition of the deposit. With two anode circuits—one copper, one tin—the control of the copper is not tied to the tin as with alloy anodes.

White Alloys

Speculum metal is a white copper-base alloy containing tin and zinc. A very hard alloy is formed by the compound Cu_3Sn, which contains 38% tin. Alloys with approximately this amount of tin or with small additions of zinc or zinc substituted for a part of the tin are hard and white. Copper-tin or copper-tin-zinc alloys that are deposited with these characteristics are considered to be speculum alloy. Bath 3 is a good example of a speculum bath. Tin may be deposited up to 60% and still retain the speculum characteristics although the alloys do become softer with increasing tin beyond the composition of Cu_3Sn. This bath is used to produce semi-bright, easily buffable alloys that are applied to indoor decorative and functional hardware.

Bright Alloy Plating

A copper-tin-zinc alloy plating process developed by the Westinghouse Electric and Manufacturing Co.[5] is a good illustration of the advantages to be gained by alloy plating.

The plating solution is prepared with copper cyanide, zinc cyanide, sodium stannate, sodium hydroxide, sodium cyanide, and an addition agent, the total metal concentration being less than 1 ounce per gallon.

At a current density of 10 to 15 amperes per square foot the analysis of the deposit is approximately 56% copper, 28% tin, and 16% zinc.

The deposit is corrosion resistant and wear resistant, similarly to speculum metal, and in addition is bright.

Anodes of the same composition as the deposit are used. The anode efficiency is 100%, while the cathode efficiency is only 35%. Therefore, a number of insoluble steel anodes must be used together with the soluble anodes to maintain bath balance.

The control of such a bath looks difficult when one considers that three metals are deposited at the same time from a bath of low metal content and at wide differences in anode and cathode efficiencies. Nevertheless, a number of tanks were operated and controlled successfully over a period of years.

A knowledge of the requirements of an alkaline tin bath and cyanide copper and zinc baths is not as helpful here as in the case of bronze plating, although some of the principles of operation of the component baths will remain unchanged. A high pH must be maintained and the condition of the anodes must be watched as in an alkaline tin bath. The cyanide content must be frequently adjusted to control the copper-plating rate.

The bath is operated more as a strike bath than as most plating baths and it has the very high throwing power associated with strike baths. The low metal content is an advantage in that the bath is readily affected by solution of metal from the anodes. Since the anodes are of the composition desired in the deposit, they aid in bath control.

An 0.2 mil thick deposit gives very good protection against corrosion, and it is harder and more wear resistant than nickel. The deposit is nonmagnetic and is easily soldered.

Preparation of the basis metal prior to plating includes cleaning and etching steps, and although the deposit may be applied directly over steel, it is recommended that copper be deposited first.

Steel Anodes

Several interesting bronze baths, used for many years, were developed by the Special Chemicals Corporation.[6] These baths demonstrate that practical plating can be done by the use of insoluble anodes and addition of chemicals to maintain bath control. The baths are maintained with replenishing salts and by adjustment of the pH and free cyanide.

A hard, white tarnish-resistant alloy, maintained by this means, was used for decorative purposes and as an undercoating for silver. It contained 62% copper, 13% zinc, and 25% tin. Because the alloy was solderable and nonmagnetic it was used on the inside of electrical instruments. An alloy with the color of 14 carat gold is obtained with 69% copper, 29.6% zinc, and 1.4% tin. A low brass color is obtained with 89% copper, 9% zinc, and 2% tin.

Applications

There is a great deal of versatility in the bronze alloys. A wide range of properties is achieved by variation of the tin and by deposition of zinc in addition. Substantial changes in hardness, tarnish resistance, solderability, buffability, ductility, and color are realized by variation of the alloy.

Small amounts of tin co-deposited from a copper cyanide bath have a marked addition-agent effect. Copper deposited from a conventional cyanide bath becomes rough after a few thousandths of an inch is deposited. When tin is co-deposited, 0.020 inch and greater is deposited with ease. Unfortunately, the strike qualities of a copper bath are lost when tin is co-deposited. Strikes or preplates are required to bond bronze deposits.

Bronze can satisfy one of the roles of nickel when used as an undercoating for other metals and can substitute for silver as well as chromium when used as a decorative coating.

Bronze can be applied purely for engineering purposes as a bearing material. It will function as a process coating when used as a stop-off during nitriding. Good solderability is an additional asset when speculum metals are used in electronic applications.

Bronze plating demonstrates the potential of changes in properties through the use of alloy plating. A look at the properties reveals that the alloy composition can be shifted to favor hardness, color, or other properties. A look at the baths reveals that the alloy composition can be shifted by manipulation of metal content, temperature and current density. Further changes are easily made by regulation of free cyanide, free caustic, and substitution of potassium salts for sodium salts.

It is not necessarily easy to change a bronze bath in a single desired direction. Nor is it possible to favor a particular property without changing others, but there is a great versatility available.

REFERENCES

1. A. K. Graham, *Proc. Am. Electroplaters' Soc.* **35,** 143 (1948).
2. N. Hall, *Metal Finishing Guidebook*, p. 198 (1965).
3. W. H. Safranek, W. J. Neill and D. E. Seelback, *Steel*, p. 102 (Dec. 21, 1953).
4. U.S. Patent 2,511,395.
5. M. B. Diggin and G. W. Jernstedt, *Proc. Am. Electroplaters' Soc.* **32,** 247 (1944).
6. U.S. Patents 2,079,842; 2,198,365.

13. CADMIUM PLATING

Cadmium is a corrodible metal; however, in mildly corrosive environments it is resistant, sacrificial to steel, and competitive with zinc. It is more resistant to salt and alkaline environments than zinc and the corrosion products that form are less voluminous than those formed on zinc.

Cadmium is present in small amounts in zinc ores and production of cadmium is dependent on the production of zinc. Because of this circumstance it is economically restricted as a by-product of zinc production. This situation keeps the price of cadmium high by comparison and at times it becomes limited as a strategic metal.

Cadmium is applied to ferrous products when it is desired to provide protection with a thin sacrificial coating. Coatings in the range of 0.2 to 1.0 mil are used to protect steel indoors, to resist occasional salt or alkali, or to provide limited protection outdoors. The coatings are used in mild marine environments or where a coating must be sacrificial as well as solderable. Cadmium is highly toxic and should never be used on food handling equipment. Also exposure to fumes during heating or soldering must be avoided.

Chemistry

Cadmium dissolves readily in dilute acids and forms soluble chloride, sulfate, and fluoborate salts. Cadmium oxide is insoluble and is not dissolved in alkaline solutions as is zinc oxide. CdO and slightly soluble $Cd(CN)_2$ can be dissolved in cyanide solutions to form complex cyanides such as $Na_2Cd(CN)_4$.

Baths

Cadmium is deposited principally from the cyanide baths and to a lesser extent from the acid baths. The cyanide bath has the advantage as the better known bath with good throwing power. Cadmium cannot be bonded to steel from cyanide solutions although adhesion is good and generally satisfactory. When a bonded deposit is desired it can be accomplished with an acid bath.

Cyanide Baths

The baths of Soderberg and Westbrook[1] present composition ranges that have been used as a basis for present day formulations:

	Still Plating		*Barrel Plating*	
	g/l	*oz/gal*	*g/l*	*oz/gal*
Cadmium oxide	23–39	3.0–5.2	20–32	2.7–4.3
Cadmium metal	19–34	2.5–4.5	17–28	2.3–3.8
Sodium cyanide	86–131	11.5–17.5	66–110	8.8–15.0
Addition agents		as required		

NaCN/Cd Ratio

It has been found by experiment that a bath can be operated within a wide total concentration range if the sodium cyanide/cadmium metal ratio is held to

$$NaCN/Cd = 3.75 \pm 0.4$$

When this is done in practice the quality of the plating will be satisfactory; however, the limiting current density will be higher at higher metal concentrations, so that occasional evaluations and changes in the operating current density will be necessary to use the bath at optimum efficiency.

Plating Within Limits

An alternate means of operating a bath depends on close limits selected from the broader acceptable range:

	g/l	*oz/gal*
Cd	27.0– 33.0	3.6– 4.4
NaCN	105.0–120.0	14.0–16.0
NaOH	11.3–22.5	1.3– 3.0
Na_2CO_3	15.0– 45.0	2.0– 6.0

Plating characteristics are better controlled when a bath is held within a narrow range. The NaCN:Cd ratio need not be calculated since the tight plating limits will keep it within reasonable bounds: it will vary from 3.2 to 4.4. A shift in plating range as detected by plating tests is more meaningful with restricted limits.

The bath can be made up with cadmium oxide, sodium cyanide, and sodium carbonate; or with cadmium cyanide, sodium cyanide, and sodium carbonate.

Cadmium oxide reacts to produce sodium hydroxide:

$$CdO + 2Na^{+} + H_2O = Cd^{++} + 2\ NaOH$$

Cadmium cyanide is assumed to combine with sodium cyanide to form a complex:

$$Cd(CN)_2 + 2NaCN = Na_2Cd(CN)_4$$

This assumption is of little value although it does provide a means of estimating free cyanide.

Total cyanide is the entire amount of cyanide in the bath expressed as sodium cyanide.

Combined cyanide is the amount of cyanide that forms the assumed complex, expressed as sodium cyanide.

Free cyanide is the total sodium cyanide minus combined sodium cyanide.

Free sodium cyanide aids anode corrosion and it is helpful to know how this value can be adjusted when an anode problem exists.

Sodium hydroxide increases conductivity and so does free cyanide to a lesser extent. An increase in either of these will lower bath voltage. Sodium hydroxide is more effective and has little influence on other bath characteristics, as does free cyanide.

Sodium carbonate is not harmful to the bath in moderate amounts. Small amounts are added to the fresh bath to promote stability. Excessive amounts cause a loss in plating range.

Cadmium in the bath is held within economic limits. A lower limit is set to assure a high efficiency and a broad plating range. An upper limit is set to hold the NaCN:Cd ratio and to avoid unnecessary loss of metal by drag-out.

Barrel plating baths are generally operated at lower metal concentrations and at higher NaCN:Cd ratios. A typical barrel bath held to narrow limits is

	g/l	*oz/gal*
Cd	15–20	2.0–2.7
NaCN	70–90	9.3–12.0
NaOH	7–15	0.9–2.0
Na_2CO_3	15–60	2.0–8.0

Baths with a high metal content favor rapid plating. A formulation for such a bath is:

	g/l	*oz/gal*
Cd	30.0–36.0	4.0–4.8
NaCN	112.0–143.0	15.0–19.0
NaOH	11.3–22.5	1.5–3.0
Na_2CO_3	15.0–45.0	2.0–6.0

High throwing power is at times important to plate on extreme shapes or to cover with thin deposits. A bath with lower cathode efficiency, brought about by low metal content and high free cyanide, favors throwing power:

	g/l	*oz/gal*
Cd	15–20	2.0–2.6
NaCN	120–135	16.0–18.0
NaOH	7–15	1.0–2.0
Na_2CO_3	15–45	2.0–6.0

A compromise of any of the above baths can be made up by arbitrarily selecting a bath that overlaps the given limits. A bath also readily can be converted from one type to another by making a discard or additions.

Characteristics and Operation

Cadmium cyanide baths operate at high cathode efficiency. The common range is 90 to 95%, but the low efficiency bath will drop to 85%. The anode efficiency is essentially 100%.

Typical operating cathode current densities range from 5 to 40 amperes per square foot. Higher current densities are attainable with high metal content and with vigorous agitation.

Bath temperatures run from room temperature to 95°F. If the baths tend to heat beyond this due to the work load it is better to provide cooling than to accept the loss in plating range.

Cyanide baths may be operated in either steel or plastic-lined tanks. Steel tanks are satisfactory but plastic-lined tanks prevent stray currents and minimize build-up of ferrocyanide (due to a slow reaction between steel and cyanide).

A bath is made up by dissolving the sodium cyanide and then adding the metal compound. When a fresh bath is made the cyanide is added to the high limit and the carbonate to the low limit. Initially, the cadmium is added to the middle of the range since it may increase or decrease during the operation.

It is good practice to analyze the fresh bath and run a plating test to establish an initial reference condition.

Bath control is primarily dependent on chemical analyses and plating tests.

Brighteners extend the plating range in addition to improving the appearance and so are almost always used. Many proprietary organic brighteners are available. The bright range can be further extended by the use of nickel and cobalt salts as secondary addition agents.

High purity cadmium is used for anodes and steel anodes are often used to supplement the cadmium anodes. This is done to avoid build up in cadmium when the anode efficiency is sufficiently higher than the cathode efficiency. However, to control properly the amount of cadmium dissolved it is necessary to provide two circuits—to the two types of anodes—and control each by the use of rheostats. Steel and cadmium are also used together by the well known method of using ball anodes in steel baskets.

At high anodic current densities, cadmium anodes become coated with a white oxidation product. The anodes then give off oxygen and function as insoluble anodes. If it is necessary to plate at maximum current density and also to plate internal areas by the use of internal anodes, it is better to use steel anodes than polarized cadmium anodes.

Control

The baths are best controlled to narrow limits supplemented by plating tests. The bath limits are controlled with the aid of analyses for cadmium, total cyanide, carbonate, and hydroxide. Brighteners are controlled by a plating test. The minimum concentration that will produce the broadest plating range is the proper amount to use. When a secondary metallic addition agent is added it is well to control it by analysis. This is desirable for two reasons. First, it eliminates the problem of determining the optimum addition of two agents by a series of plating tests. Second, it avoids the possibility of adding an excess of an agent that is not easily removed.

Filtration is generally not necessary in these baths. However, rough deposits are sometimes caused by foreign materials entering the bath or by the addition of excessive addition agents. Circumstances may require periodic or even continuous filtration.

When carbonate becomes excessive, the bath should be treated to bring the carbonate within the desired range. This can be done by cooling to about 32°F or by treatment with calcium sulfate. The calcium sulfate treatment, however, removes carbonate only at the expense of building up sulfate.

Plating Troubles

The cadmium cyanide bath is susceptible to inferior plating when metallic impurities build up in the bath. Troublesome, noble metal impurities can be introduced from the anodes or from direct contamination. High purity anodes are used to avoid the first source. Good rinsing practice will help to reduce the second source. Lead, copper, silver, antimony, thallium, and tin will co-deposit and produce slate colored or streaked plating. These can be detected in the low current range of a plating test. The metals can be removed by low current density electrolysis or by treatment of the bath with cadmium sponge, followed by filtration.

Dull cadmium deposits are usually an indication of an out-of-tolerance condition. They are avoided by maintaining limits and by plating tests to determine the cause and correction.

Burnt deposits are caused by high current areas due to improper current control.

Pitted deposits are often an indication of inadequate cleaning or pickling. It can also be associated with a low caustic content.

Noble metal impurities can immersion-plate on the anodes even while they are working, so that an anode sludge can form from the bath as easily as from the metal itself.

A high bath voltage will result if the anodes are allowed to polarize.

Conventional cadmium cyanide plating baths will embrittle high strength steels of Rockwell C 40 or greater hardness by co-deposition and entrapment of hydrogen.

Acid Cadmium Plating

Good cadmium deposits can be produced from a cadmium fluoborate bath. The bath is quite stable and the cathode efficiency is essentially 100%. Acid cadmium baths have overcome difficulties experienced in plating cast irons with the cyanide bath. Also, acid baths offer a means of producing bonded cadmium deposits.

Bath

	g/l	*oz/gal*
Cadmium fluoborate	180–240	24.0–32.0
Ammonium fluoborate	60–90	8.0–12.0
Boric acid	20–27	2.7–3.6
Addition agent	by plating test	
Fluoboric acid	to pH of 2.5 to 3.5	

The bath is operated at 30 to 60 amperes per square foot at temperatures of 70 to 100°F. With mechanical or air agitation much higher current densities are possible.

The throwing power of this bath is poor as compared to the cyanide bath.

The fluoborate bath is used for barrel plating more than for still plating.

Hydrogen Embrittlement

The cadmium fluoborate bath is preferred by some for cadmium plating of high strength steels. It puts less hydrogen into the steel than the lower efficiency cyanide baths. To avoid embrittlement, however, the preparation of the steel must also keep the introduction of hydrogen to a minimum. This is done by restricting acid treatment to a very short room temperature dip. In addition, the parts are heated after plating, usually to 375 ± 25°F for 3 hours. The conventional cadmium cyanide bath is not satisfactory as a low embrittlement bath since this cadmium deposit acts as a barrier to excess hydrogen released during processing. Special cadmium cyanide baths such as the cadmium-titanium bath have proved successful to overcome the embrittlement problem.[2]

REFERENCES

1. G. Soderberg and L. R. Westbrook, *Trans. Electrochem. Soc.*, **80,** 431 (1941).
2. D. M. Erlwein and R. E. Short, *Metal Progress*, **87,** (2), 93 (1965).

14. CHROMATE COATINGS

Fundamentally, zinc is a more reactive metal than steel. This generalization accounts for the fact that zinc sacrificially protects ferrous substrates. However, zinc does not corrode as rapidly as steel, under atmospheric conditions, and is therefore a practical sacrificial coating. Yet it does corrode, and the corrosion products are unattractive, in addition to which they sometimes interfere with the function of the part. The "white rust" that forms on zinc can greatly be delayed by covering the surface with a chemically applied chromate coating.

Treatment with an activated chromate solution will promote formation of chromate coatings on reactive zinc, cadmium, and aluminum surfaces.

The Cronak process[1] (developed by the New Jersey Zinc Company) produced protective yellow-to-brown films on zinc and cadmium. The solution contained sodium dichromate acidified with sulfuric acid. When the metal reacts with the acidified solution, an insoluble chromate gel is formed that adheres to the surface. The coating contains trivalent chromium formed by reduction and hexavalent chromium and may be reagarded as a hydrated chromium chromate. The soft gel is easily damaged when fresh but hardens and becomes tough after aging for one day. The aged film is resistant to mild wear but heavy abrasive wear will scratch the bare metal right through the film.

Chromate coatings are widely used to enhance the corrosion resistance of zinc, cadmium, and aluminum. To a lesser extent they are used as a paint base. Still less they are used to retain a conductive surface with better reliability than uncoated metal. The films can be produced in clear, iridescent, yellow, brown and olive-drab colors. On some industrial parts these typical colors have become associated with the hardware, and as such, the coatings may be regarded as decorative. Some of the coatings may be bleached and some may be dyed. Many proprietary processes are available to supply a variety of coatings. Different coatings are used for each need. Corrosion resistant coatings are heavy, thinner coatings are for paint base, and clear and conductive coatings are even thinner.

Protective value of the film is dependent on the inhibitive action of slightly soluble chromate supplied to a corrodible surface in the presence of moisture.

Chromate coatings are best applied to zinc and cadmium as a part of the plating cycle: (1) plate, (2) rinse, (3) chromate, (4) rinse. (5) dry.

Drying temperatures must be kept below 150°F to avoid loss of integrity of the film. The hexavalent chromium is converted to an inactive form by temperatures above 150°F.

If chromate is applied to surfaces that are not freshly plated, or to zinc alloys, the metal should be cleaned, then activated by an acid dip prior to chromating. The coating may be bleached, dyed, or both while in process and before drying. A hot rinse frequently is inserted as a final step to facilitate drying. The time in this rinse is kept to a minimum to avoid excessive softenting and leaching of the film. Some processes include a final hot rinse, at temperatures up to boiling, as a bleach.

Bath control is accomplished by additions of chromate chemicals and acid or activator, based on determinations of hexavalent chromium and pH.

With experience, the integrity of the film can be estimated from the color, uniformity, and gloss. Minimum quality is sometimes assured by specification of a minimum time without formation of white corrosion products in a salt spray. Military specifications for zinc and cadmium specify 96 hours.[2]

Low pH chromate treatments are employed for bright dipping of zinc and cadmium. These processes produce a clear bright film or a yellow film that is bleached in hot water, warm alkali, or proprietary bleaches. The coatings are often lacquered to improve corrosion and wear resistance.

Aluminum Treatments

Yellow and brown films can be formed on aluminum by immersion, spray, or swab with activated chromate solutions. The coatings have characteristics and properties quite similar to those formed on zinc and cadmium.

The Alodine process (developed by the American Chemical Paint Company in 1945[3]) is more versatile than the chromating process. These coatings were formed in acidified chromate-phosphate-fluoride solutions. They are chromate-phosphate coatings, capable of heavier coverings. Coatings of 0.1 to 0.4 mils are formed in a few minutes in cold or warm solutions. These coatings appreciably extend the life in natural corrosive environments. Thin iridescent coatings are applied for a paint base.

The coating may be regarded as a chromium-aluminum-phosphate. It is thin and hard, iridescent, or light green in color, depending on the thickness. Heavier and darker green coatings are produced by more recently developed and improved processes.

REFERENCES

1. E. A. Anderson, *Proc. Am. Electroplaters' Soc.*, **30,** 6 (1943).
2. Military Specifications *QQ-Z*-325 *Zinc Plating; QQ-P*-416 *Cadmium Plating.*
3. U.S. Patent 2,438,877.

15. CHROMIUM PLATING

Commercial chromium plating stems from the work of Fink and co-workers as described in two early patents.[1] These baths are quite simple, consisting of a solution of chromic acid and a small amount of sulfuric acid. Plating with these baths has been highly successful, and the baths, still widely used today, are best designated as "conventional."

Chromium is an attractive bluish-white metal with excellent resistance to staining. Corrosion resistance of the metal is dependent on a very thin oxide film that retains the metallic luster and accounts for the passive state of the metal. The metal is also very hard and therefore resistant to abrasion and wear. Because of these properties, a very thin coating of chromium provides a durable as well as an attractive finish. These qualities account for "chrome" as a standard of excellence.

Chemistry

The conventional chromium plating bath contains chromic acid and sulfuric acid in the ratio of 100:1. The chromic acid is decomposed electrolytically to form chromium on the work and to produce water and oxygen. Roughly one fifth of the current is used to produce chromium. The remaining part is expended to decompose water into its elements.

Chromic acid is formed from chromic trioxide and water:

$$CrO_3 + H_2O = H_2CrO_4$$

The role of the two acids in the bath may be represented by the following changes that take place during plating:

$$100H_2CrO_4 + H_2SO_4 = 100Cr + 100H_2O + 150O_2 + H_2SO_4$$

These elementary reactions do not, of course, take place in this simple way, but in the language of chemistry they do represent what happens—namely, that the chromic acid is decomposed to the metal plus harmless by-products and that the sulfuric acid remains to be used over. It is a unique fact that the chromic acid bath can continuously be operated with insoluble anodes without a build-up of decomposition products.

The sulfuric acid that remains in the bath is the "catalyst" that is essential if any plating is to take place. It must be present within very specific limits to control the quantity and the character of the deposit.

Conventional Baths

The historical chromium baths are the 33- and 50-ounce baths referred to as the "dilute" and "concentrated" baths.

When limits are assigned to these baths the practical operating requirements become:

Table 8

	Dilute Bath		*Conc. Bath*	
	oz/gal	*g/l*	*oz/gal*	*g/l*
CrO_3	30–33	225–250	48–53	360–400
H_2SO_4	0.30–0.33	2.2–2.5	0.48–0.53	3.6–4.0
°Bé at 60°F	19.7–21.3		28.9–31.1	

The cathode efficiency and the plating rates are higher in the dilute bath. The concentrated bath, on the other hand, has better throwing power and can be operated at lower voltage. Dilute baths are favored by lower drag-out, wheras concentrated baths are less responsive to chemical changes. It is desirable to have a little of the advantages of each, and this has led to the conclusion that an "average bath is a good compromise.

Table 9 Average Bath

	oz/gal	*g/l*
CrO_3	36–40	270–300
H_2SO_4	0.36–0.40	2.7–3.0
°Bé at 60°F	23.1–25.0	

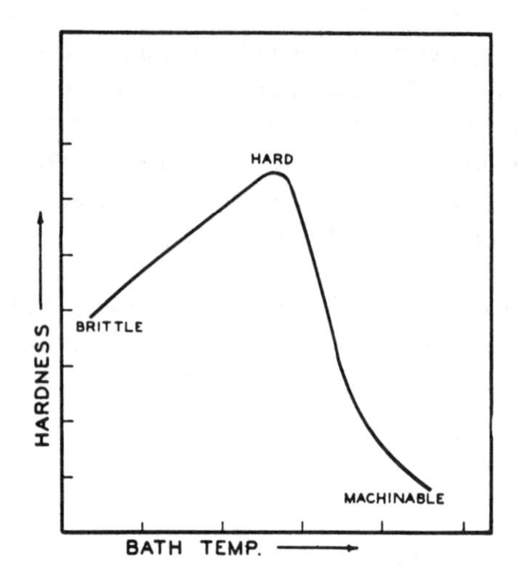

Fig. 8 Chromium hardness vs. bath temperature.

Hard and Bright Plating

In engineering applications, reference is often made to hard or industrial chromium and also to bright chromium. In ordinary plating practice, hard and bright chromium are one and the same, for bright deposits are hard and, in addition, the control of the hardness of the deposit is too difficult for the average plater.

In fact, it may be said that there is only one conventional chromium bath. By changing the concentration, the temperature, and the current density, the characteristics of the deposit can be changed. The effect of bath temperature on the hardness of the chromium deposit is shown in Figure 8.

Operation

The chromic acid bath is a good example of how a bath is operated by a knowledge of its simple variables. If the following characteristics and variables are kept in mind, it will be found that the optimum operating conditions can be obtained.

1. The bath operates at a high current density.
2. If the current density is increased, the temperature must be also increased.
3. If the current density is increased, the plating rate will be also increased.
4. If the current density is increased, a higher voltage is required.
5. If the plate is bright, it is also hard.
6. The throwing power is poor.

The operation of the bath can be best illustrated by examples.

Let us assume that it is desired to plate 0.1 mil of chromium on steel and that the plating rate is not important. In a 33 ounce per gallon bath, a bright plate will be obtained at 0.75 ampere per square inch and it will take about 12 minutes to deposit 0.1 mil. If the bath has a number of parts in it so that the total cathode area is large, the bath temperature will increase because of the resistance of the electrolyte. If the bath gradually increases in temperature to 105°F, it may be necessary to increase the current to 1 to 1.5 amperes per square inch and to cut the plating time to 8 minutes to maintain a bright plate. This will cause further heating and require further increase in current, so that with continuous operation the bath may come to equilibrium at approximately 120°F, at a current density of 2 amperes per square inch and a plating time of 6 minutes.

From this example it can be seen that conditions will have to be continuously changed if appreciable total current is used. It is poor practice to continuously change the current density and the plating time. Therefore, for the above application, it would be best to heat the tank to 120°F before starting operation and to cut the heat input as heat is supplied from the electrolysis. If a large area of work is to be plated, it is necessary to have both heating and cooling coils. Tables are available for the plating rates of chromium baths and they should be used as a guide.[2] However, the tables were determined under conditions of uniform current density and give only approximate values for irregular cathodes. In practical applications, a prepared rack is placed in the chromium-plating tank at the proper temperature and the current density is increased until the plate becomes bright. The total current or voltage is noted and this or some slightly higher value is used for operation.

Shop records or a log are more helpful to chromium plating than to any other common bath. The current to be applied to a part often cannot acurately be estimated. The amount of current is applied that will do the job within the limits imposed by the shape of the part and the anode arrangement. After the part is plated the thickness is measured and the plating rate is recorded on the shop card. When no record exists the plating time is estimated from known approximate rates.

It is a rule of thumb that good conventional deposits are produced when the plating rate is one mil per hour. Approximate data are given in Table 10.

Table 10 Approximate Current Density to Plate One Mil per Hour

°F	*Dilute Bath*		*Conc. Bath*	
	Asi	*Asf*	*Asi*	*Asf*
100	1.7	250	2.1	300
120	2.4	350	2.8	400
140	3.0	430	3.3	480

Continuous quality chromium plating takes the bath characteristics into account. The outstanding characteristic of the chromium bath is its limited plating range. The plating range is limited because the covering power and throwing power are low. Also, the plating range shifts with change in temperature or change in current density. However, a shift due to change in temperature can be compensated for by a change in current density.

The shift in plating range with temperature change is illustrated in Figure 9. Since the current does not distribute well over the cathode,

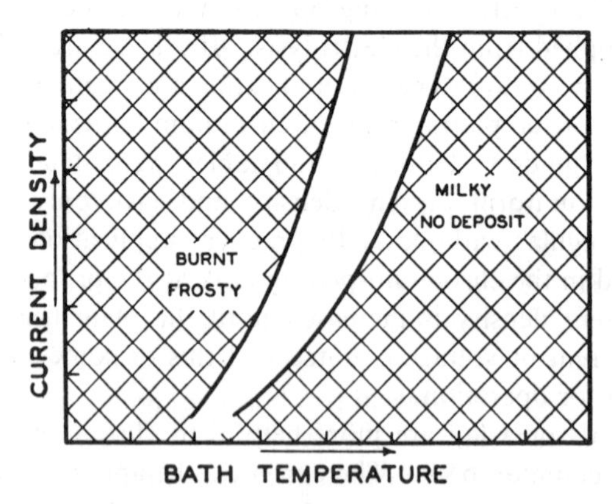

Fig. 9 Chromium plating range: current density vs. bath temperature.

it is essential to use an average current density well within the range.

The operation of a chromium bath is relatively simple, but as in all baths there are certain characteristics that must be watched. The solution, in addition to being a plating bath, has fair cleaning action and can also be used to etch steel if the steel is treated anodically in the bath. Not only will it etch steel on anodic treatment, but the etching action is more effective on many alloy steels than with the more common methods of etching. In the common etching methods, using hydrochloric or sulfuric acid, a smut is often left on alloy steels which, if not removed, may cause poor adhesion between the steel and the electrodeposited overlay.

Since the chromium bath has the combined features of cleaning, etching, and plating baths, it may be used as a complete plating line with the addition of nothing more than a rinse tank. To illustrate the simplicity of operation, it is well to consider how some manufacturing plants use the bath for salvage of tools. The steel to be plated is immersed in the bath and treated anodically until the desired etch is obtained. The current is then reversed and the steel treated cathodically until the desired thickness of plate is obtained. Chromic acid baths have been used in this manner for limited plating in the vicinity of the plant tool-salvage department. This practice, however, is not recommended for continuous plating or for large-scale plating.

Control

It is a relatively simple matter to control a chromium bath. The minimum basic requirements are (1) a measurement of the gravity, (2) analysis for sulfuric acid, (3) a plating test.

A gravity measurement readily reveals the approximate concentration of chromic acid. Unless an unusual quantity of impurities is present this is a sufficient indication. It is possible to control the bath by means of a hydrometer and a plating test, the latter being used to detect the optimum catalyst concentration. It is better to analyze chemically for sulfuric acid and use a plating test as an overall check when the gravity and the sulfate are within the prescribed limits. The plating test will then reveal troublesome impurities or a deviation from optimum plating.

The concentration of sulfuric acid in the chromic acid bath is very low. Its effect on the conductivity of the bath is negligible. How-

Table 11 Specific Gravity and Chromic Acid Content*

Specific Gravity at 15°/4°C	*Baumé at 60°F*	*CrO_3 oz/gal*
1.13	16.7	24.8
1.14	17.8	26.8
1.15	18.9	28.8
1.16	20.0	30.6
1.17	21.1	32.6
1.18	22.1	34.4
1.19	23.2	36.4
1.20	24.2	38.6
1.21	25.2	40.3
1.22	26.2	42.3
1.23	27.1	44.2
1.24	28.1	46.2
1.25	29.0	48.2
1.26	29.9	50.2
1.27	30.8	52.2
1.28	31.7	54.5
1.29	32.6	56.5
1.30	33.5	58.7
1.31	34.3	60.7

* The presence of trivalent chromium and iron will affect these readings.

ever, it is very important that the recommended ratio of CrO_3 to H_2SO_4 be maintained. The ratio may vary from 50 : 1 to 250 : 1, but in most cases satisfactory results are obtained if the ratio is held at approximately 100 : 1. It is imperative that the catalyst ratio be held at an optimum since the plating range of the bath is narrow even under the best conditions.

The gravities of solutions of CrO_3 in water are given in Table 11.

Anodes

Lead, lead-tin, or lead-antimony are used in practice. These lead or lead-alloy anodes are insoluble in the bath. However, during operation, the surface of the anodes becomes covered with a film of lead peroxide that will oxidize any trivalent chromium present.

Iron anodes are occasionally used in applications where the weight or physical characteristics of the lead-alloy anodes are objectionable. The build-up of iron in the bath, however, will eventually necessitate dumping of the bath.

Hexavalent chromium is continuously reduced to trivalent chromium at the cathode. When the concentration of trivalent chromium becomes excessive, the deposit turns gray. The trivalent chromium is reoxidized at the anode. This reoxidation keeps the trivalent concentration low when the anode area is large with respect to the cathode and when the anodes are operating properly. In use, lead anodes form a film of lead peroxide that promotes reoxidation. If the lead peroxide becomes too thick it will interfere with the anodic current. Also, the anodes can become passive on standing. In either instance the anodes are restored to proper operation by cleaning. Removal of the anodes from the tank when it is not in use is troublesome but helpful if the tank is out of service for long periods of time.

Occasional analyses for trivalent chromium are advised as a guide to the efficiency of reoxidation.

The major problem in chromium plating is current distribution, which requires special attention to racking and to anode positioning in order to compensate for the poor throwing power of the bath. If complicated shapes are plated, a complex anode must be made by bending lead strips or wires to conform approximately to the contours of the cathode.

Other Chromium Baths

There are a number of successful and popular variations of the chromic acid bath that are quite widely used. The foremost of these is the self-regulating bath known as SRHS (self-regulating, high-speed).[3] Other variations of the bath will produce "porous," "black," "ductile," or "crackfree" chromium. From the terminology alone it is readily obvious that the appearance and the properties of chromium can greatly be changed.

SRHS Baths

These baths contain sulfate catalyst and a second catalyst containing fluoride in some form. Chemicals are added to the bath that are of such a nature that the bath is saturated with compounds that limit the sulfate and the fluoride automatically at desired concentrations. This constitutes self-regulation and does away with the need to control the catalysts by analysis. The cathode efficiency is higher in this bath and the bath has a wider plating range.

When a bath that is saturated with catalyst is heated, the solubility of the catalysts will increase. In order to maintain the chromate-to-catalyst ratio it is necessary to add chromic acid. At lower temperatures a less concentrated bath is used. The fluoride in the bath promotes etching that results in several effects on metal in the bath. It will more easily activate passive nickel but on the other hand it promotes etching of the lead anodes.

Crack-Free Bath

A self-regulating bath is available that will produce satiny crack-free deposits.[4] These deposits are buffable and offer improved corrosion resistance.

Porous Chromium

Porous chromium[5] is an oil-retaining modification that has greatly extended the life of cylinder bores and other lubricated wearing parts. The surface is produced by plating under conditions that favor nodular plating. This surface is then subjected to etching in a chromic acid solution similar to the plating bath, or by other chemical or electrochemical means. The part is then honed to produce the oil-retaining surface. A similar result can be obtained by roughening the surface prior to plating.

Black Chromium

A "black" chromium deposit is produced in a bath containing chromic and acetic acids.[6] The deposit comprises a mixture of chromium and oxides of chromium that have a nonreflective gray color.

Preparation of the Basis Metal

Typical low carbon steels can be prepared for plating by conventional cleaning and pickling practices. Alloy steels that are used to make heat-treatable steel parts cannot properly be prepared by these same procedures. It is fortunate that most steels can be prepared by anodic etching in the bath itself or a similar solution. Prior to hard chromium plating, parts are anodically alkaline cleaned and then anodically chromic-acid etched for ¼ minute up to 15 minutes depending on the type of steel and the heat treatment. Copper brass and bronze do not present any particular problems. Zinc and aluminum are generally preplated with other metals as is steel that is to be decoratively plated.

Construction Materials

Chromium-plating tanks may be either lead, brick, or plastic lined. Antimonial lead or lead-tin/alloy linings are more resistant to corrosion from the bath than chemical lead. The lead or lead-alloy linings are susceptible to the effects of stray currents that may cause chromium to deposit on the lead, may rob the cathode of current, may allow current to flow to areas on the cathode where it is not desired, and may accelerate corrosion of the lining. However, if proper precautions are taken, the lead linings are satisfactory and will last for years. Brick lining has the advantage of eliminating stray currents, but it is more difficult and expensive to install than lead.

Plastic lined tanks provide a nonconductive covering for the steel tank at a much lower cost. Fiberglass and polyvinyl chloride are used. Uncoated steel tanks also are used. They will contaminate the solution but they have been used in some instances for years.

Plating Thickness

Chromium plating can be classified by thickness more than by any other property. Decorative chromium is used principally to provide a stain resistant finish over nickel. This can be done with 0.01 to 0.02 mils. Deposits less than 0.02 mils are porous, whereas thicker deposits are cracked. Proprietary baths have been developed that produce a "microcracked" chromium that provides added protection when thicknesses are increased up to 0.05 mils.

Hard chromium is plated from 0.05 mil up to 20 mils. Thin deposits are used on cutting tools. The thin chromium increases tool life by providing a low friction surface. Coatings greater than 0.05 mil produce a dulling effect on the cutting edge. Parts that are subject to light wear such as gages are plated to size with 0.1 to 0.2 mil. Parts subject to heavier wear, such as shafts, are generally plated 5 to 6 mils and ground to 2 to 3 mils. Much heavier coatings are used to salvage mismachined parts (if necessary).

Bath Troubles

The major troubles in the chromic acid plating bath are hinged around the poor throwing power and the limited plating range of the bath. It is frequently necessary to try and try again. An irregular shaped part may require robbing as well as conforming anodes. Under these conditions it is difficult to know what portion of the current is reaching the area to be plated. It is then necessary to pick a probable total current, plate, measure the thickness, then estimate the proper current and time for the next try. If the current is too high or robbing is insufficient the corners will burn, tree, or form nodules. If current does not reach recessed areas at a sufficient density the plate will not cover. These problems are solved principally by ingenuity in rack design and occasionally by redesign of the part.

REFERENCES

1. U.S. Patents 1,581,188 and 1,802,463.
2. G. Dubpernell, *Trans. Electrochem Soc.* **80,** 589 (1941).
3. J. E. Stareck et. al. *Proc. Am. Electroplaters' Soc.*, **37,** 31 (1950).
4. R. Dow and J. E. Stareck, *Proc. Am. Electroplaters' Soc.*, **40,** 53 (1953).
5. H. van der Horst, *Proc. Am. Electroplaters' Soc.* **31,** 56 (1943).
6. E. A. Ollard, *J. Electrodepositors Tech. Soc.* **12,** 33 (1936).

16. ACID COPPER PLATING

Copper is the most abundant of the noble metals. It is corrosion resistant, ductile, and highly conductive. It resists attack by nonoxidizing acids but is readily dissolved in oxidizing acids. Although it is corrosion resistant, it is highly susceptible to staining and therefore limited as a decorative coating. It is also limited as a result of forming a noble-metal base-metal couple that invites corrosion when it is used as a coating for steel aluminum or zinc.

Acid copper plating is about as old as the art of plating itself. A look at the properties of copper sulfate readily suggests why this salt has successfully been used over the entire period. Copper sulfate (more properly, cupric sulfate) is the most common cupric salt. A very stable solution is formed by dissolving this salt in water acidified with sulfuric acid. In such a solution, copper is dissolved anodically at an anode efficiency of 100%.

$$Cu = Cu^{++} + 2e$$

Due to the fact that copper salts are easily reduced, the same reaction proceeds in the reverse direction at the cathode with the same efficiency. Copper sulfate is appreciably soluble, and since it is a noble metal it will deposit in preference to more base metals such as iron and zinc that are likely to contaminate the solution. In addition to these qualities it happens that metal of good quality can be deposited from solutions of copper sulfate and sulfuric acid without the aid of addition agents.

Acid Versus Cyanide Baths

An acid plating bath such as the acid copper bath is the opposite in many respects to a cyanide bath such as the cadmium cyanide bath. A comparison of the general characteristics of the two types of baths

Table 12 Comparison of Acid and Cyanide Plating Baths

Characteristic	*Acid Copper*	*Cyanide Cadmium*
Plating rate	almost unlimited	limited
Electrode efficiencies	essentially 100%	variable
Control	simple	somewhat complex
Throwing power	poor	good
Covering power	poor	good
Availability of metal	high	low

is given in Table 12.

The comparison of the two baths, as given in Table 12 will generally hold true for all acid baths and cyanide baths. The main reason for the difference between these two types of baths is that acid baths contain simple ions whereas cyanide baths contain complex ions. More accurately, in acid baths a fair percentage of the metal is present as simple ions, while in cyanide baths only a very small percentage of the metal is present as simple ions and a large percentage as complex ions.

Acids Used

Any acid that will form a soluble copper salt is a potential plating candidate. Many acids have been tried with some success, such as acetic, hydrochloric, nitric, and fluosilicic. Sulfamic acid has been quite successful. Sulfuric and fluoboric acids are most commonly used because of marked accomplishment with simple solutions.

Sulfuric Acid Bath

The following limits in the bath formula have been suggested[1]:

	g/l	*oz/gal*
Copper sulfate (crystals)	150–250	20–33
Sulfuric acid	45–100	6–13

The composition of the bath may be varied, depending on the plating rate and the type of deposit desired.

The copper sulfate is the source of the metal ions and a high metal content makes possible a high plating rate. The sulfuric acid is also essential and the solubility of the copper sulfate is limited by the presence of the acid. It is not advisable to operate too near the solubility limit since any increase in salt content or a decrease in bath temperature will result in formation of copper sulfate crystals.

Sulfuric acid provides good bath conductivity and prevents the formation of basic copper compounds. The sulfuric acid also decreases the amount of copper ions present for the same amount of total copper in the electrolyte. As the copper-ion content is decreased, the crystal size of the deposit is also decreased.

The characteristics of the bath and the properties of the deposit will be more consistent with narrow limits. The following is typical of an easily controlled bath that will produce consistently:

	g/l	*oz/gal*
$CuSO_4 \cdot 5H_2O$	200–250	27–33
H_2SO_4	45– 75	6–10

Bath Preparation

The bath is made up in plastic-lined or ceramic tanks. Special precautions regarding the addition of chemicals are not too important except that the sulfuric acid should be added to a large volume of water in order to prevent spattering and overheating. The copper sulfate crystal contains five molecules of water of crystallization and has the formula $CuSO_4.5H_2O$. There is no danger of caking of the crystal, although considerable agitation may be required to bring about solution.

Bath Operation

The copper sulfate bath will produce good deposits at room temperature. Current densities of 20 to 50 amperes per square foot are common. With mild agitation the plating rate may be doubled and with vigorous work movement it may be increased ten times. The deposits are sound although they have a grainy appearance. These deposits are soft, buffable, and easily formed, but they also readily tree at edges and irregularities. The grain can be refined by the addition of glue, peptone, molasses, and many other organics. Simple, single addition agents are used to control treeing. Other agents, such as thiourea, promote bright deposits and proprietary agents are available that offer full bright deposits.

Since the cathode efficiency is 100% and as cathode polarization in the bath is low, the acid copper bath has a narrow plating range. This means that both covering power and throwing power are low and that particular attention should be paid to the positioning of the anodes. It also means that unless addition agents are used treeing will take place in high current-density areas and that even if addition agents are used, the deposit will be heavy in high current-density areas (corners and edges).

The high cathode efficiency and the fact that the limiting current density can be greatly increased by the use of agitation make it possible to obtain very high plating rates. However, heavy deposits will become rough after depositing a few mils unless addition agents are used. If it is permissible to remove part of the deposit mechanically, heavy deposits may be produced that will be sound, but they will be too rough to be made smooth by buffing.

During operation of the bath, the copper sulfate content increases and the sulfuric acid content decreases. The anode efficiency is really greater than 100%, calculated at a copper valence of two, because part of the copper goes into solution at a valence of one. If drag-out is high, the accumulation of copper may be partially taken care of, but some lead anodes may be introduced into the tank to reach a solution balance. Whether the solution is balanced or not, some sulfuric acid will have to be added at intervals to compensate for the drag-out of sulfate content.

By agitating the bath and by increasing the temperature, higher current densities may be used—with a resulting increase in the plating rate.

Relatively soft deposits can be obtained if these are required by operations following the plating. However, the conditions that produce soft deposits usually favor crystalline deposits and treeing in high-current-density areas. Harder and finely crystalline deposits are favored by the following conditions: (1) low copper concentration, (2) high acid concentration, (3) high current density within the plating range, (4) low solution temperature, (5) low agitation rate, (6) use of addition agents,

The first five conditions are those that do not favor replenishment of copper ions to the cathode in the immediate vicinity of the cathode. The sixth condition covers those addition agents that interfere with crystal growth and produce fine crystals.

If rapid plating in desired, those conditions should be used that favor rapid replenishment of the copper ions in the immediate vicinity of the cathode.

Anodes

Rolled copper anodes are preferred for acid copper baths, but cast copper anodes and electrolytic sheet are used.

As a general rule, anode current densities should be about the same as the cathode current densities. At still plating conditions, the anode current density may be up to 50 amperes per square foot. With agitation the limiting current density is increased about in proportion to the increase at the cathode.

Films form on the anodes that consist of copper oxide, copper particles, and a sludge of metal impurities that are not solubilized by the electrolytic action. Anode sludge can migrate to the work or be carried to it when the bath is agitated. When necessary this trouble is kept in check by anode bagging and regular filtration. The trouble is avoided or at least greatly reduced by using high purity anodes, rolled anodes, and cast anodes containing small amounts of phosphorous.

Insoluble lead alloy anodes can be used when a fixed anode geometry is essential to control current distribution. The bath, of course, must either be rejuvenated or frequently adjusted when insoluble anodes are used.

Control

The control of the bath is simple. The gravity is measured and a titration made for the acid content. The gravity measurement determines the total content of copper sulfate crystals and sulfuric acid, while the acid content is calculated from the titration. The copper sulfate crystal content is then determined by subtraction of the acid content from the total.

The following equation expresses the relationship between the two chemicals present and the specific gravity (with sufficient accuracy for control purposes):

$$\text{sp. gr. at } 70^\circ\text{F} = 1.000 + 0.00460 \times (\text{oz/gal of } H_2SO_4 + CuSO_4 \cdot 5H_2O)$$

Fluoboric Acid Bath

The characteristics of the fluoboric acid bath are very similar to the sulfate bath with the added advantage that the metal concentration can be doubled.

Commercial copper fluoborate concentrates are available that can be diluted and adjusted within the following broad operating range:

	g/l	*oz/gal*
Copper fluoborate	225–450	30–60
Fluoboric acid	15–30	2–4
Boric acid	15–30	2–4

An average bath with assigned limits is as follows:

	g/l	*oz/gal*
$Cu(BF_4)_2$	330–360	44.0–48.0
HBF_4	20–25	2.7–3.3
H_3BO_3	20–25	2.7–3.3
Sp.gr (70°F)	1.27–1.29	
pH	0.6–0.9	

A dilute 30-ounce fluoborate bath will have operating characteristics similar to the sulfate bath. This bath is operated at a pH of 1.5. At 60 ounces the bath is maintained at a higher acidity in the pH vicinity of 0.4. Warm high-concentration baths can be operated with mild agitation at current densities up to 400. These baths can be operated without addition agents, although the addition agents used in the sulfate baths will generally modify the deposits.

Pyrophosphate Bath

This bath is made up with copper pyrophosphate, potassium pyrophosphate, and ammonia. These salts form an alkaline pyrophosphate complex that enables operation in a pH range between 8 and 9. Although a complex is formed, the characteristics of the bath are more like those of the acid baths than of the cyanide baths.

Bright plates are produced by the use of addition agents, and since the bath is close to neutrality it minimizes the tendency to attack substrates that is associated both with the strongly acid baths and the alkaline-cyanide baths. Because of this very slight alkalinity the bath is preferred for plating on zinc, aluminum, and plastics.

There are many applications where this bath offers distinct advantages, and at the same time it can be used as an alternate to the acid baths.

Preparation of the Cathode

If acid copper is deposited on a noble metal sùch as silver, all that is required to obtain bond is to clean and etch the basis metal prior to plating. The choice of etch depends on the basis metal; a nitric acid etch or anodic treatment in a cyanide solution would provide a satisfactory etch for silver or copper.

If acid copper is deposited on such base metals as nickel, iron, or zinc, then bond cannot be obtained directly. Since the copper is noble in acid solutions, it will deposit by chemical displacement, and even though the current is flowing, some immersion-plating will take place. The immersion plate will be loose, i.e. nonadherent. It is therefore necessary to first plate the base metal with a noble metal using a bath from which the metal will not immersion-plate. A cyanide copper bath is such a bath. The accepted procedure is to plate the base metal with cyanide copper and then deposit acid copper over the cyanide copper. Copper will not immersion-plate on steel from a cyanide bath because steel is a noble metal in a cyanide bath. Therefore, to deposit copper on steel, the usual steps are taken and a cyanide copper-plating step is inserted following the etch and preceeding the acid copper plating. When a cyanide copper bath is used in this fashion, it is known as a strike and more will be said about it in the next chapter.

Bath Troubles

These copper baths are easy to operate and control, but they suffer from the poor throwing power that is typical of acid baths. The deposits also have a strong tendency to build up at edges, making it essential to control them by anode arrangement, careful racking, and possibly robbing or shadowing. In some instances it is advisable to redesign parts to avoid sharp edges, corners, and recesses. The use of addition agents will aid in suppressing build-up of current-concentration areas.

Decomposition of addition agents and accumulation of other organic materials can cause nonuniform, rough, brittle, or burned deposits.

Anode sludge and other particles that become suspended in the bath will promote rough plating.

Roughness that is associated with suspended particles is solved by filtration, whereas roughness or burning associated with organics requires activated carbon treatment and filtering.

Chlorides pit copper and can cause trouble both at the anode and at the cathode.

Lead will cause brittleness in fluoboric acid baths. It can be precipitated by the addition of sulfuric acid.

Influence of Basis Metal

Some interesting work has been done on the study of the initial crystalline form of acid copper deposits on basis metals prepared by various pretreatments.[2] These studies are a key to the adherence of the deposit.

If acid copper is deposited on clean and etched copper, the deposit will tend to repeat the structure of the basis metal at and near the bond line. This indicates that the deposit must be close enough to the basis metal to be influenced by the atomic forces. However, if the basis metal

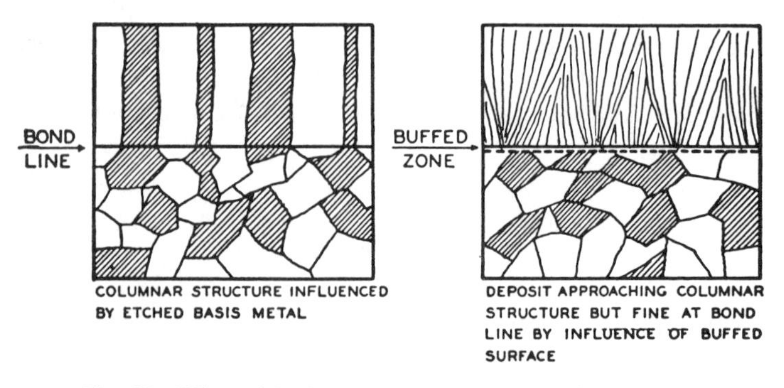

Fig. 10 Effect of basis metal on the structure of the deposit.

is buffed, it will not influence the crystalline form of the deposit in the same manner.

The crystal structure of a copper deposit may even be influenced by the crystal structure of a basis metal other than copper if the crystal system of the basis metal is similar to that of copper.

Figure 10 is a diagrammatic representation of the influence of etched basis metal on the deposit.

Applications

Acid copper baths are used for electroforming, electrotyping, and electroplating. If acid copper is to be used for electroforming, complete lack of bond is essential. The mold may be metallic or nonmetallic. If the mold is metallic, it is cleaned and dipped in a solution of 1 part of light motor oil to 3 parts of naphtha. After evaporation of the naphtha, a light film of oil remains on the mold and this serves as a parting compound over which the copper may be deposited and from which it may be stripped.

If the mold is a nonconductor, such as wax, rubber, plaster, glass, or plastic, then a conductive film must be applied. If the surface is porous, it must be sealed with shellac, varnish, or lacquer. A conductive film of graphite or bronze powder may then be applied, or a conductive film of silver may be applied by the reduction process used for silvering glass. This film may be obtained by immersion in a silvering solution or by simultaneously spraying with one solution containing silver nitrate and another solution containing the reducing agent.

By the electroforming process it is possible to make sheets, plates, tubes, pipe fittings, bells for musical instruments, tanks, or any shape where a master can be made and an appropriate electrode system can be economically arranged. Thus, the electroforming process should be used for making the article from the pure metal. An outstanding feature of the process is that the reproduction of detail is extremely fine. This is apparent from the fact that acid copper is used in making phonograph records, starting with an original wax master.

The surface of an electroformed copper article can be plated with nickel or other metal or combination of metals, if so desired.

Copper plating has a long history of service as an undercoat for nickel. Great quantities of copper-nickel-chromium trim have been produced for automotive and other decorative applications. Many prefer nickel alone as an undercoat to avoid the potential troubles from a copper base-metal cell. Others prefer the economy of a soft buffable undercoat and rely on quality overcoats to completely cover the copper. Cyanide copper or acid copper over a copper strike are used for undercoats.

REFERENCES

1. J. H. Winkler, *Trans. Electrochem. Soc.*, **80,** 523 (1941).
2. W. Blum and H. S. Rawdon, *Trans. Electrochem. Soc.*, **44,** 305 (1935); A. W. Hothersall, *Trans. Faraday Soc.*, **31,** 1242 (1935).

17. COPPER CYANIDE BATHS

Cyanide solutions provide a medium for direct deposition of copper on base metals, particularly steel and zinc. Copper will readily deposit on passage of current but it will not immersion-plate on steel. When the steel is properly prepared by cleaning and etching, the deposit will bond. A thin copper deposit is often used as a bonding layer, or a "strike," in preparation for further plating.

Good copper deposits are easily produced with little tendency to roughness, treeing, or edge build-up. Cyanide copper baths overcome the major disadvantages of the acid baths. On the other hand they do not offer the advantages of control simplicity, stability, economy, and ease of producing heavy deposits. It is important to point out again that the cyanide copper baths and the acid copper baths usually do not compete. When the advantages of one are needed the other is a poor substitute.

The copper cyanide baths may be divided into three types:

1. Plain copper cyanide baths.
2. Rochelle copper cyanide baths.
3. High-efficiency copper cyanide baths.

All contain copper cyanide and sodium or potassium cyanide.

The baths may be applied in three ways: (1) strike plating, (2) underplating, (3) finish plating.

These three categories do not correspond with the three types of baths. A plain copper bath or a Rochelle bath can be used as a strike, whereas the high-efficiency bath requires a strike as a preplate prior to plating on base metals. Any of the three may be used as an undercoat and each could be a final coating.

The plain copper bath is restricted to thin deposits, usually from .02 to .10 mils. The Rochelle bath is preferred up to 0.5 mils and the high-efficiency bath up to 2.0 mils. Greater thicknesses are not attained with conventional plating procedures.

Chemistry of the Baths

The copper in the acid baths is bivalent (Cu^{++}) and requires 2 faradays to deposit 1 mol. In the cyanide baths the copper is monovalent (Cu^{+}) but can only be deposited with 1 faraday from high-efficiency baths. The Rochelle baths are much less efficient and a large part of the current produces hydrogen.

The chemistry of the copper cyanide baths is not simple since more than one complex is formed. However, for bath formulation and control purposes it is safe to assume the following reaction:

$$2NaCN + CuCN = Na_2Cu(CN)_3$$

Thus, it is assumed that two moles of sodium cyanide combine with 1 mole of copper cyanide; and it will be found on titration for free cyanide that all of the sodium cyanide except 2 moles will combine with silver nitrate used for the determination. Since this method is used for control and since it measures all the sodium cyanide present but the two combined moles, the assumption is valid on a control basis.

Influence of Free Cyanide

The behavior of the cyanide baths is understandable if the effect of free cyanide is considered (all the sodium cyanide in excess of 2 moles). When the free cyanide content is zero, a substantial cathode efficiency is obtained even at low temperatures. However, as free cyanide is added, the cathode efficiency drops and plating practically stops at 4 moles of total sodium cyanide for 1 mole of copper cyanide (or at 2 moles of free sodium cyanide).

The plating rate can be increased by increasing the temperature. But there is more required of a plating bath than a substantial plating rate. The deposit must be satisfactory and the anodes must be soluble.

When free cyanide is low, the anodes polarize. A black film is formed and the anodes give off oxygen gas and do not supply copper to the solution. During this process, copper is removed from the solution (at the cathode) and free cyanide builds up with a resulting decrease in the plating rate. Eventually, the increase in free cyanide will aid anode corrosion, but control under such conditions is difficult. Because of these difficult control conditions, the simple low-temperature baths are not widely used. However, they may be used as a strike where plating is not continuous.

For volume production plating, the Rochelle copper bath is generally used. Here, the higher temperatures and Rochelle salt aid anode corrosion and make a higher plating rate possible.

For more rapid plating, the high-efficiency copper baths are used in which the temperature is high, the metal content high, and the free cyanide low. These baths require several proprietary addition agents for successful operation.

Plain Cyanide Baths

The plain cyanide bath may be controlled by adjusting the free cyanide to obtain the desired plating rate. Figure 11 illustrates how the cathode efficiency is affected by the free cyanide content.

As the free cyanide is increased, the plating rate drops. If copper cyanide is added to the bath the metal content will be increased and at the same time the free cyanide decreased. Both of these changes will result in an increase in the plating rate. Therefore, control is critical. Also, if soluble anodes are used, they should be adjusted so that balance is maintained. If the metal builds up in the bath too rapidly, steel anodes may be used for part of the anode area.

The plating rate may be increased by increasing the current density, but even though the plating rate increases, the cathode efficiency drops rapidly and at high current density, very little is gained. Also, if high current density is necessary, a large anode area may be required to

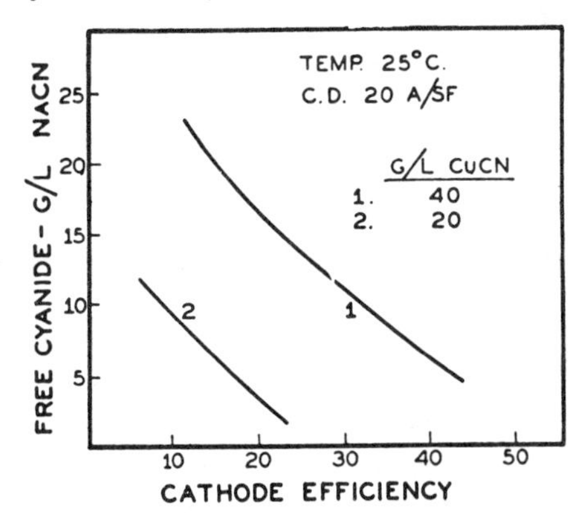

Fig. 11 Copper cyanide cathode efficiency vs. free cyanide.

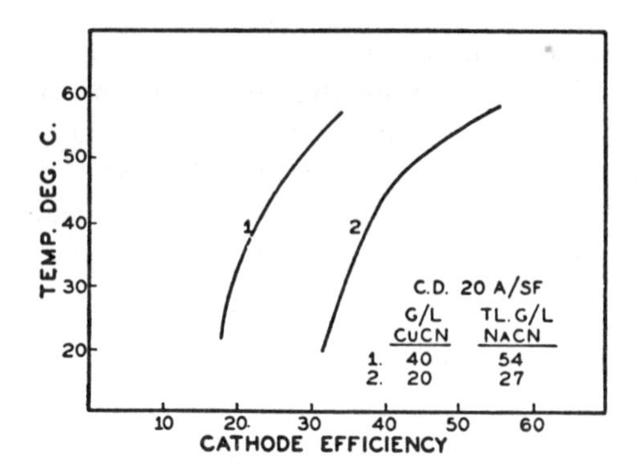

Fig. 12 Copper cyanide cathode efficiency vs. temperature.

prevent polarization of the anodes.

The plain cyanide bath is difficult to control continuously at room temperature because the bath is very sensitive to changes when it is operated at low efficiency. It is best to increase the temperature and thus favor more stable conditions. An increase in temperature increases the cathode efficiency and also reduces anode polarization. Figure 12 shows how the cathode efficiency increases with increase in temperature.

If the plain copper cyanide bath is to be used to produce thin deposits, it is well to measure the cathode efficiency frequently and to balance the bath by adjusting the anode area, the temperature and the current density, and by the use of some steel anodes.

If a plain copper cyanide bath is to be used as a strike, it is well to make occasional measurements of cathode efficiency. The bath may be varied widely to meet different requirements. The anode efficiency will not be as important as for continuous plating because the strike will only be used at intervals. Such a bath may even be controlled by the use of steel anodes and the addition of chemicals to maintain the plating rate.

Very thin deposits of the order of 0.01 to 0.02 mil are often all that are required for striking, and this thickness is easily obtained with a striking time of 1 minute in cold baths. However, it is sometimes desired that the strike also have some cleaning power. In this case, it is desirable to decrease the cathode efficiency by increasing the free cyanide, increasing the current density, and raising the pH—all of which increase the cleaning power of the bath. It is also desirable to operate at elevated temperatures. High current density, high temperatures, and high pH cause the strike to act as an electrolytic cleaner. The pH may be increased by adding sodium hydroxide or trisodium phosphate.

The following formula is recommended for a plain copper cyanide bath:

	g/l	*oz/gal*
CuCN	19–26	2.5–3.5
Free NaCN	5–10	0.7–1.4
Na_2CO_3	15–60	2.0–8.0
pH	11–12.2	
Temp., °F	70–90 *or* 100–120	
Current density	5–10 *or* 10–20	

Rochelle Bath

The Rochelle bath can function as a strike, as a copper plating bath, or as both. The Rochelle salt in the bath aids anode corrosion and produces a finer-grained structure. By operation at higher temperatures the limitations of low efficiency and the control difficulties of the plain cyanide bath are overcome.

The following formula is recommended for the Rochelle bath[1]:

	g/l	*oz/gal*
Copper cyanide	19–45	2.5–6
Sodium cyanide	26–53	3.5–7
Sodium carbonate	15–60	2.0–8
Rochelle salt	30–60	4.0–8
Free sodium cyanide	15–30	2.0–4
current density	20–70 amp/sq ft	
temperature	130–160°F	
pH	12.2–12.8	
anode current density	10–30 amp/sq ft	
cathode efficiency	30–60%	

Narrower limits can be selected and held when it is desired to restrict the range of the cathode efficiency. The cathode efficiency responds to temperature, current density, and free cyanide effects in a manner similar to the plain cyanide bath. The pH is held within limits to control the appearance of the deposit. However, changes do not take place rapidly in the bath, and with some adjustments a balanced bath can be obtained.

High-Efficiency Copper Cyanide Baths

The high-efficiency copper cyanide baths carry to an extreme the conditions that will produce high cathode efficiency. These conditions are: (1) high metal content, (2) low free cyanide, (3) high temperature, (4) agitation.

In addition, the bath is operated at a relatively high caustic content and contains proprietary addition agents.

The following bath limits are recommended for the sodium bath[2]:

	g/l	*oz/gal*
Copper cyanide	90.0 –150.0	12.0 –20.0
Sodium cyanide	100.0 –170.0	13.5 –22.5
Free sodium cyanide	3.75– 11.25	0.5 – 1.5
Sodium hydroxide	22.5 – 37.5	3.0 – 5.0
Brightener	11.1 – 18.7	1.5 – 2.5
Antipit agent	1.2 – 1.8	0.15– 0.25
current density	10–100 amp/sq ft	
temperature	168–185°F	
cathode agitation	2–15 ft/min	
anode current density	5–30 amp/sq ft	

The anode and cathode efficiencies in this bath are both essentially 100%. Thus, a balanced bath is easily maintained. In addition, the plating rate is high and the bath is relatively stable, considering the high temperatures employed. The bath gradually builds up in carbonate due to chemical decomposition, but the carbonate can be more easily removed by crystallization from this type of bath than from the baths of lower metal content. If a steel tank is used, the bath may also build up in ferrocyanide, but this does no harm and may be removed by crystallization along with the carbonate.

The disadvantages of the bath are:

1. A strike must be used prior to plating to obtain bond.
2. Cathode agitation is essential to maintain the high plating rate.
3. Addition-agent control is required.
4. Good rinsing is required.
5. Frequent or constant filtration is recommended.

A high-efficiency bath may also be prepared using potassium salts. The potassium bath is similar to the sodium bath except that less metal is required to obtain a high cathode efficiency. Carbonate cannot be crystallized from the potassium bath, but a greater amount can be tolerated and it may occur that the decomposition rate of the bath will be low enough so that removal of carbonate will be taken care of by drag-out.

If the high-efficiency bath is properly controlled, semibright deposits will be obtained that are a good base for deposition of bright nickel.

Contaminants

Cyanide baths may become contaminated with foreign metals or organic materials. These contaminants can be removed by the use of low-current-density electrolysis and treatment with activated carbon. Small-scale tests will indicate the proper treatment.

Bath Preparation

In preparing a cyanide copper bath, the sodium cyanide should be dissolved first. The copper cyanide will then be soluble in the sodium cyanide solution. After the cyanides are dissolved, the other chemicals may be added. Steel tanks may be used, but rubber- or plastic-lined tanks are preferred to prevent the formation of ferrocyanide and to prevent stray currents.

Operation and Control

In any of the cyanide baths, except strike baths that are occasionally used, an adjustment of conditions should be made to obtain balance so that the anodes are not excessively polarized and that metal is entering the solution at the same rate at which it is being removed.

Such a cathode-current density should be used that good deposits can be obtained at higher and lower current densities. In other words, the average cathode-current density should be somewhere in the middle of the plating range.

By comparison of the three types of copper cyanide baths it will be seen that conditions can be changed so that one of the baths approaches the performance of another. For instance, if the Rochelle bath is operated at room temperature, its characteristics will be much like that of the plain bath. On the other hand, if it is operated at high temperature and low free cyanide content, it will approach the performance of the high-efficiency bath. Thus, a quality deposit can be obtained over a wide range of temperatures and with a wide variation in plating rate. However, the quality of the deposit is not the only measure of a good plating bath. If conditions are greatly different from those recommended, trouble may develop with bath balance and with anode corrosion. If difficulty is experienced, and if bath decomposition is high, frequent additions of chemicals may be required. If excessive additions are required, bath control will be too difficult for continuous plating. In experimenting with variations of a formula, a good log will be helpful to anticipate whether continuous control is feasible. If this is done, there is no reason why experimentation cannot be carried out in fullscale plating.

In all of the baths, analysis for copper and free cyanide is essential for control of the metal content and the plating rate. The pH is controlled in the Rochelle bath, but at the higher pH of the high-efficiency bath an analysis is made for sodium hydroxide. Also, in the high-efficiency bath the free cyanide content must be determined cold by the use of cracked ice (at 4°C or less). Addition agents are controlled by plating-range or by special tests recommended by the supplier.

Bath Characteristics

One characteristic of the copper cyanide baths is that heavy deposits cannot be obtained. Good deposits are possible up to 2 mils, but beyond this thickness the deposit becomes rough. Next to limiting thickness, the baths are notable for the drop in cathode efficiency with increase in free cyanide. Cathode efficiency can be increased by the use of low current density and high temperature. Agitation will increase the limiting current density for high-efficiency baths, but if agitation is employed, frequent filtration is required. When a bath is agitated, suspended particles come in contact with the cathode more easily and cause rough deposits.

Anodes

Rolled and annealed anodes generally work well in the copper cyanide baths, although electrolytic anodes are recommended for the high-efficiency baths. If the free cyanide is low, the temperature low, or the anode-current density high, the anodes will polarize and lead to increase in free-cyanide content, loss in metal content, aud acceleration of bath decomposition.

Preparation for Plating

The basis metal is cleaned and etched by the usual procedures prior to plating. Copper may then be deposited on the basis metal and good bond will be obtained. If a good copper deposit is obtained, almost any metal may be deposited over it. This is the manner in which the low-efficiency baths function as a strike, and low-efficiency plating is required even prior to deposition from a high-efficiency copper cyanide bath.

Heavy Deposits

Commercial experience has shown that the 2-mil thickness limitation can be overcome by periodic plating and deplating.

REFERENCES

1. A. K. Graham and H. J. Read, *Trans. Electrochem. Soc.*, **80,** 341 (1941).
2. H. L. Berner and C. J. Wernlund, *Trans. Electrochem. Soc.*, **80,** 355 (1941).

18. IRON PLATING

Iron deposits differ from others in that they have no decorative or protective value. Iron is deposited for low cost and to take advantage of the properties of the metal. Electroplating provides high purity metal that can be rendered soft and ductile by annealing or can be hardened by the processes of cyaniding, nitriding, or carburizing. By choice of baths and variations in the plating conditions, the hardness of the metal can be changed considerably but to a lesser degree than by metallurgical means. Due to the purity the deposited metal also provides good magnetic properties.

Iron plating is applied for electrotyping, electroforming of shapes, reinforcing of nickel electroforms, and salvage of worn or mismachined parts.

Iron Chloride Baths

Very good iron deposits are produced with ferrous chloride solutions. These baths are highly corrosive and must be operated hot. Control of the baths requires much more attention than most plating baths, but when these baths are properly monitored, excellent, thick deposits will result.

Kasper[1] demonstrated that soft, ductile deposits can be produced from hot, concentrated solutions of ferrous chloride with a small excess of hydrochloric acid. Later, Stoddard[2] showed that a finer-grained deposit could be formed by the addition of manganese chloride to the bath. The following range is recommended:

	g/l		*oz/gal*
$FeCl_2 \cdot 4H_2O$	250–300		33.0–40.0
$MnCl_2 \cdot 4H_2O$	3–5		0.2– 0.5
Temp. °F		180–200	
Current density, amp/sq ft		50–75	
pH		1.5–2.0	

Control of the free hydrochloric acid is absolutely essential to the operation of an iron chloride bath. On the other hand, operation becomes routine when proper attention is given to this requirement. The free acid is defined by the pH or it is better determined by titration. Control to a range of .05 to .10 oz/gal HCl (.01 to .02 N) will keep the bath operating properly. When the acidity is below this range the bath oxidizes and produces ferric iron, which in turn promotes rough plating. When the acidity is above this range the cathode efficiency is low. In a hot bath it takes frequent adjustments to maintain the range. The excess acid is used by chemical attack of the anodes and the deposit. In a relatively short time the free acid is sufficiently low to be out of the operating range. Actually, the bath can be operated with free acid above the recommended range, but unfortunately the acid is used up much more rapidly when the acidity is high. The best compromise to this control problem is to add acid continuously at a rate that will maintain the desired free acid.

Commercial ferrous chloride contains small amounts of ferric salts that are detrimental to the bath. The ferric iron can be reduced to ferrous iron by working the bath in the presence of a small excess of hydrochloric acid. When small amounts of ferric iron are present (from a fresh make-up or from oxidation on standing), it is easily reduced by working the bath after an addition of 1 oz/gal of HCl. The acid is rapidly used to form ferrous iron at operating temperatures. Complete reduction can be recognized by a change in color of the bath from a brownish green to a pure green. When larger amounts of ferric iron are present, due to long standing without free acid present, larger amounts of acid may be added. At boiling temperatures and with large anode areas present, this acid can be made to work rapidly. After the bath is reduced it is maintained by keeping the small required acid concentration present. Liberties can be taken with the acid treatment when adjusting the bath, but control of the acid must be strict when the bath is operated.

High-purity iron anodes must be used. If the carbon content of the iron is much over 0.02% carbides will accumulate on the surface. The carbides will float off as a smut, become suspended in the bath, and eventually cause a rough plate. Even when high-purity iron anodes are used it is recommended that the anodes be bagged.

The ferrous chloride concentration of the bath can be allowed to increase, due to working, with little change in characteristics, up to 66 oz/gal. Higher concentrations of ferrous chloride are beneficial to increase conductivity.

The bath can be operated at lower temperatures but it is necessary to reduce the current density to retain ductility. Operation below 160°F is not recommended.

Iron Sulfate Baths

Operation at low temperature overcomes the disadvantages of the iron chloride baths. At the high temperatures necessary the ferrous chloride baths are very corrosive and present the problem of close monitoring to sustain control. Good iron deposits can be formed from ferrous sulfate solutions at lower temperatures. The deposits are more brittle than those from the chloride baths and the bath is not as easy to keep reduced and free of ferric iron because the bath is less active. Thus, bath activity is associated with advantages and disadvantages and indeed the baths might be generally classified as low- and high-activity baths.

MacFadyen[3] recommended the following bath:

	g/l	*oz/gal*
$FeSO_4 \cdot (NH_4)_2SO_4 \cdot 6H_2O$	350	47
current density	65 amp/sq ft	
temperature	140°F	
pH	6	

In general, the maintenance of the bath is similar to that of the chloride bath. The bath must be kept reduced by continuous electrolysis or by occasional treatment with iron and excess acid followed by adjustment to the proper pH. A pH of 6 is maintained by keeping a suspension of ferrous carbonate and powdered charcoal in the bath.

This bath is much more subject to pitting than the chloride bath and precipitates form easily at a pH of 6.

The tendency to pit and the tendency to form precipitates can be reduced by operating at higher temperature and at lower pH. Under these conditions, the deposit will be more ductile.

Sulfate-Chloride Baths

Schaffert and Gonser[4] recommend a sulfate-chloride electrolyte that was particularly developed for electrotyping and surfacing of stereotypes:

	g/l	*oz/gal*
$FeSO_4 \cdot 7H_2O$	250	33.0
$FeCl_2 \cdot 4H_2O$	42	5.6
NH_4Cl	20	2.7
current density	50–100 amp/sq ft	
temperature	100–105°F	
pH	4.5–6.0	

Reducing treatment, pH adjustment, filtering, and carbon treatment are required during preparation of the bath.

A consideration of various formulas for iron baths will show that iron may be deposited from sulfate or chloride solutions where the acidity is low. Additions of ammonium, sodium, magnesium, and other salts may be made that will alter the characteristics of the baths. By changing the pH, the temperature, and the current density, the hardness of the deposit can be changed. As the bath composition is changed, the control problem also changes.

The recommended formulas should be used unless a new plating problem arises that warrants considerable experimentation. The low-temperature baths are attractive because the heating and ventilating problems are simple, but the hot chloride bath is easy to operate continuously if the proper pH is maintained and the recommended practice is followed.

REFERENCES

1. C. Kasper, *J. Research Natl. Bur. Standards*, **18,** 535 (1937).
2. W. B. Stoddard, *Trans. Electrochem. Soc.*, **84,** 305 (1943).
3. W. A. MacFadyen, *Trans. Faraday Soc.*, **45,** 455 (1924).
4. R. M. Schaffert and B. M. Gonser, *Trans. Electrochem. Soc.*, **84,** 319 (1943).

19. LEAD PLATING

Lead plating is not applicable to decorative applications since the coating is not pleasing and is quite susceptible to wear. The coating is limited largely to protection of the substrate from corrosion in specific environments. The soft, ductile, low-melting metal is also useful as a bearing surface.

Lead fluosilicate is used as a low cost practical electrolyte for electrolytic refining. Lead fluoborate is quite prominent for electroplating and lead sulfamate offers a second choice.

Fluoboric acid can be made by dissolving boric acid in hydrofluoric acid. This is a troublesome and somewhat hazardous procedure since the reaction takes place rather vigorously and the hydrofluoric acid can inflict poisonous burns. It is much more practical to use commercial fluoboric acid and metal fluoboric concentrates. A knowledge of the fact that the fluoboric acid is a reaction product of two other acids is, however, useful control information. The reaction conforms approximately to the equation:

$$4HF + H_3BO_3 = HBF_4 + 3HO_2$$

An excess of boric acid is added to assure complete reaction of the HF. Metal fluoborate concentrates contain free fluoboric acid as well as free boric acid.

The limits of the plating bath are not critical and the following concentrations are satisfactory:

	g/l	*oz/gal*
Lead	80–120	10–16
Free fluoboric acid	30–60	4–8
Free boric acid	7–11	1–1.5
Current density	10–50 amp/sq ft	
Temperature	70–100°F	
Glue	as required	

Bath Preparation

The bath is prepared in rubber- or plastic-lined tanks by adding the desired amounts of lead fluoborate solution and fluoboric acid to water and adjusting to volume. Usually some precipitate is formed due to small amounts of sulfates in the water. This will settle to the bottom of the tank and cause no trouble; however if the initial deposit is rough, filtering is indicated. The solution thus prepared will deposit fair lead but the coating will be crystalline, will have a strong tendency to tree, and will have very poor covering power. An addition of 0.1 ounce per gallon of glue or gelatine is added to the bath to improve these conditions. When dry glue is used it is treated with hot water to obtain a dispersion before it is added to the bath.

Bath Characteristics

The anode and cathode efficiencies are 100%, so the bath does not change readily with use. Operation of the bath is relatively simple if the common problems of poor throwing power and poor covering power are accepted. Great liberties can be taken with the chemical concentrations if desired.

Good deposits can be obtained from baths of a wide range of lead concentration, and with higher lead contents, higher current densities may be employed. At a concentration of 7 ounces lead per gallon, good deposits can be obtained at a current density of 10 amperes per square foot. At 10 ounces lead per gallon, an average current density of 20 is used, but current densities of 40 amperes per square foot are possible. If the lead content is increased to 25 ounces per gallon, the current density may be increased to 60 amperes per square foot, and with agitation even higher current densities may be used. At a current density of 60, the bath will deposit 9 mils per hour so that high plating rates are possible.

The bath may be operated with very high or very low acid content, but concentrations of 4 to 8 ounces fluoboric acid per gallon of bath are satisfactory. There is no particular reason for operating with high acid content, and control of low acid contents would be difficult, particularly since pH cannot be controlled by the ordinary electrometric method; therefore, a medium acid concentration is maintained.

Anodes

High-purity lead anodes should be used, although small amounts of the nobler metals that are found in lead will not cause trouble. Antimony and arsenic are not soluble in acids on anodic treatment in the presence of lead. These metals will collect on the surface of the anode as a sludge and eventually drop to the bottom of the tank. If the bath is agitated, the fine metal particles on the anode may become suspended and temporarily cause rough deposits, but the particles will settle and collect with other insoluble material on the bottom of the tank. The bath may occasionally be decanted and the sludge can be removed. It is worth while to remove the sludge occasionally since it is necessary to agitate the tank after additions are made and also desirable to agitate the tank occasionally.

If the tank is not agitated occasionally, it will stratify.

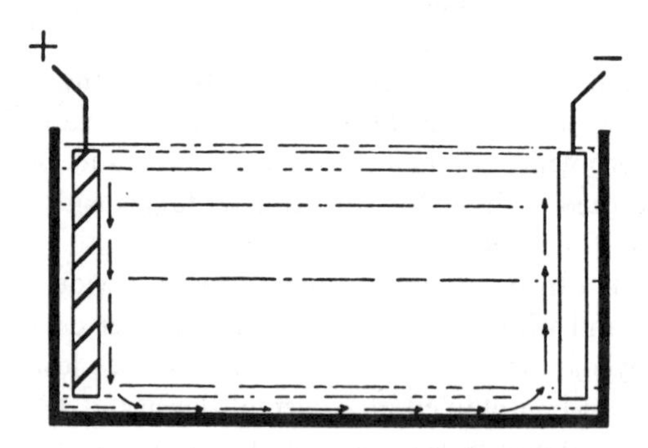

Fig. 13 Convection-stratification.

When the lead goes into solution from the anode a convection current is set up and the solution of high specific gravity near the anode stratifies along the bottom of the tank (Fig. 13). This happens more readily in lead baths than in others because solutions of high specific gravity can be formed with lead salts.

Control

The principal ingredients in the bath are easily controlled. The acid can be titrated and the metal content estimated with sufficient accuracy by gravity measurement. It is desirable to control the boric acid although the lead that is present will precipitate free fluoride as lead fluoride if it should happen that the free boric and a small amount of the combined boric would, for some reason, be consumed. This circumstance automatically avoids the possibility of accumulation of free hydrofluoric acid.

Fair deposits can be obtained without the use of an addition agent. Such deposits are crystalline and have a tendency to tree. Some users have found that they can tolerate these conditions and in fact prefer the crystalline lead that does not stain as readily as the deposit obtained with the use of glue. Also, of course, it avoids the addition agent problem.

When glue is used as addition agent it is added in small amounts as required. It is best controlled by routine plating tests and it is well to keep in mind the role that the glue plays:

When glue is added to an aqueous solution, it forms a colloid and it is the colloidal properties of the glue that account for its effect as an addition agent. Colloids carry a charge and because at least part of the colloidal glue is positively charged, it is attracted to the cathode. It is co-deposited with the lead. In the deposit, it interferes with crystal growth and aids the formation of a finely crystalline deposit. Only a part of the glue has this function so that the addition cannot be controlled by analysis for total glue content. The glue also undergoes decomposition with time so that the ratio of active to total glue is variable and is not known. If too much glue is introduced into the bath, or if the deposit becomes brittle due to changes in the glue, then it can be removed by treatment with active carbon. The active carbon will not remove all of the glue, but it will remove the portion that is active. If at times the bath requires filtration to remove suspended material, some of the active glue will be removed but it cannot be effectively removed by ordinary filtering practice. The colloid will pass through, unless the filter is packed with activated carbon.

Sulfamate Bath

The characteristics and operation of the sulfamate bath are quite similar to the fluoborate bath. Lead concentrations are similar and free acid is added to a pH of about 1.5. At lower pH values the sulfamic acid hydrolyzes and lead compounds are precipitated. However, if this condition is avoided the bath is easily controlled and maintained with the aid of addition agents.

Cathode and Deposit

When lead is deposited on steel, the steel is prepared by the usual cleaning and pickling steps. Sulfuric acid may be used for pickling, but good rinsing is important since any sulfates carried into the bath will cause formation of insoluble lead sulfate.

The covering power of the bath may be improved by the use of a copper strike prior to lead plating, but it is of no particular advantage unless a covering power problem exists. With proper cleaning and pickling, satisfactory bond to steel can be obtained without the use of a strike.

In order to protect steel from rusting, a pore-free deposit is necessary. Steel is anodic to lead, and thin deposits are porous. However, a deposit of 1 mil will give good protection unless wear is a factor. If wear is important, deposits of 5 mils thickness may be required.

20. LEAD-TIN

Lead-tin alloys can be deposited from the same baths that will deposit lead. Mostly it is required to add tin to the bath, to use lead-tin anodes, and to control the solution composition to obtain the desired Tercentage of tin in the deposit. With this approach, alloys have been deposited from perchloric acid, sulfamic acid, and fluoboric acid. The sulfamate bath has attained some prominence but the fluoborate bath is more widely accepted.

Use

Lead-tin deposits are soft and unpleasing in color. Use, therefore, is restricted to engineering or economic applications. As a corrosion resistant coating it is used on battery parts, chemical and oil containers, and sheet roofing. It is used as a running surface for bearings and as a break-in surface for machine parts. An important application of lead-tin alloys is for solder and deposits are used as such—for example, as printed circuit patterns.

Characteristics

The anode and cathode efficiencies of the fluoborate bath are essentially 100%. This aids the problem of bath balance and allows high plating rates. The most noticeable difference from the lead bath is that the change that takes place continuously now becomes important because it affects the relative deposition rates of the co-deposited metals. Closer addition agent control is essential to overcome this tendency.

The throwing and covering powers of the bath are poor but are somewhat better than lead. The addition of the tin aids these characteristics and has a grain refining effect, thus acting as a secondary addition agent.

Some bath decomposition and sludge formation takes place but it is small. Tin oxidizes and the oxidized portion becomes inert, but the ratio of stannous to stannic tin levels off at an equilibrium with use. With use the bath becomes a stable bath.

Formulation

Addition agent control becomes an essential part of the lead-tin bath. The addition agent has a marked influence on the amount of tin co-deposited. Peptone is a popular addition agent that will promote the codeposition of tin. Gelatine or glue alone are not sufficiently affective, so they may be aided by the addition of Resorcinol.

The following bath was formulated by DuRose[1]:

Bath 1

	g/l	*oz/gal*
Lead	218.0	29.00
Tin	11.0	1.50
Glue	0.8	0.10
Resorcinol	0.4	0.05

A bath of this composition will deposit an alloy containing 5% tin, sufficient to enhance considerably the corrosion resistance of the lead. The bath is operated at room temperature with lead-tin anodes and at current densities of 10 to 40 amperes per square foot. The percentage of tin in the deposit will change with the current density and with aging of the bath. As the bath ages and a part of the tin is oxidized, greater amounts of tin are added, until an equilibrium is reached.

As little as 2% tin in electroplated lead will give a more pleasing color than lead and it will not darken as readily. The corrosion protection of steel is good, and DuRose claims that an alloy deposit containing 5 to 6% tin will better protect steel than pure lead or leadtin alloys with greater percentages of tin.

Lead-tin baths need not be operated at high lead concentrations, although high concentrations will permit higher current densities. If low current densities are satisfactory, a more dilute bath may be used, or with agitation a more dilute bath may be used at higher current densities. If the bath is agitated, frequent filtration and bagging of the anodes is likely to be necessary and the agitation must be restricted either to mild or uniform to retain control over the rate of tin deposition. The concentration of chemicals, agitation, and current density are largely a matter of choice, depending on thickness desired, quality of deposit, and allowable plating time.

Bath 2 was proposed by Seabright[2] to deposit alloys of 60% tin–40% lead for soldering applications.

Bath 2

	g/l	*oz/gal*
Lead	25	3.3
Tin (total)	60	8.0
Tin (stannous)	55	7.3
Free fluoboric acid	80	10.5
Free boric acid	25	3.3
Glue	5	0.7

The total metal concentration is the major limiting factor on the overall operating characteristics of a bath. Good results can be obtained with total metal concentrations from 8 to 30 oz/gal.

Bath 3 was used to plate 8 to 10% tin for bearings.

Bath 3

	g/l	*oz/gal*
Lead	100–110	13.3–14.7
Tin (total)	15–20	20.0–27.0
Tin (stannous)	7–12	1.0–1.6
Free fluoboric acid	40–50	5.0–6.0
Free boric acid	20–30	2.7–4.0
Resorcinol	2–3	0.3–0.4
Gelatine	as needed	

Operation

Limiting current density of a bath may be increased by increasing the metal content. Doubling the metal content, however, may only increase the allowable current density by 20 to 30%, at the expense of doubling the drag-out.

The alloy anode has a self-controlling influence on the bath that tends to counteract a shift in the deposit composition that takes place as the bath naturally changes. A more dilute bath will respond to the controlling influence of the anode more readily than a concentrated bath.

Bath 2 might more properly be called a tin-lead bath since the alloy is a tin base alloy. The bath, however, operates with the characteristics of a lead bath and responds to glue as an addition agent. Tin-base alloys can be deposited with small amounts of lead, but when such is the case the baths behave like tin baths and respond better to acid tin addition agents.

The limits shown for Bath 3 have to be taken with a grain of salt because of the effects of continual decomposition of the gelatine and oxidation of the tin on the rate of deposition of the tin.

When Bath 3 is made up with 2 oz/gal of tin and 0.13 oz/gal of gelatine, a deposit of 10% tin will be obtained at a cathode current density of 20 amp/sq ft. A portion of the tin will oxidize to stannic tin with a subsequent drop in tin content in the plate. It will be necessary to add more tin to maintain the stannous tin content. After a number of weeks, with consistent operation, an equilibrium will be reached and changes due to oxidation of tin will be low. Accumulation of lead and stannous tin in the bath from the alloy anodes will tend to prevent changes in the deposit after sufficient use of the bath.

It is not possible to predict the exact bath composition to obtain deposit compositions within narrow limits. This will vary with bath conditions and with the work load. To control the bath within a narrow deposit analysis range, it is necessary to keep a log so that adjustments can be made to the desired analysis.

Gelatine and tin are the only chemicals critical to control. Resorcinol is added to minimize changes in composition with changes in current density. At low resorcinol concentrations, the tin in the deposit increases with increase in current density in the low current density range. At high resorcinol concentrations, tin in the deposit decreases with increase in current density.

With proper attention to the stannous tin and the addition agents, the operation and preparation for plating is in all respects the same as for a lead bath.

Control

The lead-tin bath contains lead fluoborate, stannous fluoborate, stannic fluoborate, fluoboric acid, boric acid, resorcinol, and gelatine. The stannous tin and gelatine are the only critical and potentially variable factors; the other bath chemicals can be controlled to well defined limits.

Fluoboric acid renders the metals soluble. Addition of acid beyond this amount—the "free" acid— increases bath conductivity. Concentration of free acid is not critical but should be controlled. Tin in the bath causes inconsistent results in the method of titration for free acid with the method ordinarily used for a lead fluoborate bath. Tin also interferes with electrometric pH measurement. The best known method for control of free acid is by colorimetric pH. A pH range of 0.5 to 1.5 is satisfactory. Free fluoboric acid may be 4 to 7 oz/gal in this range.

Boric acid is not essential to bath operation, but some free boric acid is added to make sure that all the hydrofluoric acid has been reacted. If unreacted and free hydrofluoric acid is present, a reaction will take place to precipitate lead as the fluoride. Free boric acid does no harm to the bath and the control limits of Bath 3 have been found satisfactory. Boric acid content is estimated by determining the amount required to saturate a sample portion of the bath.

All conditions being equal, the tin in the deposit is directly related to the stannous tin and independent of the stannic tin. The stannous tin is easily determined by direct iodine titration. An occasional analysis for total tin is desirable to determine when the bath has reached equilibrium. Equilibrium in Bath 3 is usually reached on a working bath when the stannous tin and stannic tin are about equal.

Resorcinol is a stable chemical and it is therefore advisable to control to definite limits. It can be determined by a bromination method.

Gelatine or glue are added to a bath to promote a fine grained deposit, to maintain the limiting current density, and to advance the deposition of tin. An analysis for gelatine is of no value; it must be controlled by observation, scheduling, and plating test.

If gelatine is not added, the deposit will be crystalline and will contain no tin. Excessive gelatin causes the deposit to become brittle and streaked. A gelatine concentration of 0.1 to 1.0 grams per liter is added to a fresh bath. Good results are achieved with daily additions of 0.01 grams per liter to a working bath. If the bath is not used for several days no harm will result but the bath will deteriorate on long standing.

With long use it becomes necessary to change the gelatine addition scheduele. If the gelatine is used or decomposed more rapidly than it is added, it is readily recognized by a drop in the percentage of tin in the deposit. If too much gelatine is present, or gelatine is coverted to a harmful form, a plating test will anticipate trouble.

After accumulated experience, lead-tin baths can be operated for long periods of time on a predetermined schedule of additions. The excess or deficiency of gelatine that can cause loss of control has been avoided by frequent small additions of gelatine (based on experience).

When glue is used, larger additions will be required but the control scheme is the same. When a schedule has been set, a tendency toward crystalline plating indicates a need for larger additions and a tendency to roughness indicates that additions should temporarily be halted. These observations, best observed by plating test, will aid in establishing the optimium addition quantity and frequency.

A plating test standard on a fresh bath is useful for control purposes. Subsequent tests that show a decrease in the plating range and a drop in the limiting current density or crystallinity at low current density indicate a need for addition agent. Streaks near the upper limit of the current range indicate excessive addition agent. Such streaks indicate that trouble will develop if further addition agent is added or if the bath is not used. Corrective steps can be deduced from testing on small portions of a bath followed by further plating tests. Electrolysis may cure the trouble. At times, an addition of gelatine may help even though this is not indicated, but this is best determined on a test scale. If the bath is producing rough brittle or cracked deposits on the work, then it is likely that activated carbon treatment and filtration are in order. This also can be established by small scale testing. If large or unusual changes in the bath are made, to bring it within limits, an analysis for tin in the deposit should be made since a change is likely.

The key to successful lead-tin control is to keep within limits, to keep a close continual watch on the addition agent, and to watch for a drift in the stannous tin limits to maintain the tin in the deposit.

Copper and Antimony Additions

Co-deposition of copper or antimony will harden the deposit. These metals are noble to lead and tin in acid solutions and will deposit in direct proportion to the concentration in the bath. They also have a grain refining and a slight addition agent effect. Concentrations of 1 to 3 grams per liter will produce deposits with about 1 to 3% of the metal in the deposit at a current density of 20 in Bath 3. These lead-tin-copper and lead-tin-antimony alloys are used for bearing surfaces.

REFERENCES

1. A. H. DuRose, *Trans. Electrochem. Soc.*, **89,** 417 (1946).
2. L. H. Seabright, *Metal Progress*, Oct. 1949, p. 509.

21. NICKEL PLATING

Nickel plating is sufficiently extensive to constitute an industry by itself. The known applications are too numerous to list and the unknown applications are no doubt even more numerous. A great number of patents have been issued and 211 references cited in an article by Pinner, Knapp, and Diggin[1] attest to the extent of the litera ture.

The word to describe nickel is "versatile." The metal is attractive, corrosion resistant, and has good abrasion resistance. It has both decorative and engineering value. The baths, too, are versatile so that the metal can be deposited hard, bright, matte, or ductile.[2] Nine recognized baths are presented in Table 13. These baths can be made up from two or more of the four basic chemicals. Bath No. 1 is a flash plating bath that is used only where thin, brittle deposits and a low plating rate can economically be justified. Bath No. 2 is recommended for plating wax or lead molds in support of electrotyping applications.

The popular Watts-type baths (Nos. 3, 4, and 5) produce good quality, ductile deposits. They are general-purpose baths with good anode efficiency, fast plating rates, and freedom from peeling and

Table 13 Compositions of Nickel Baths

	oz per gal						
Bath No.	$NiCl_2 \cdot 6H_2O$	$NiSO_4 \cdot 6H_2O$	NH_4Cl	H_3BO_4	*pH*	*Bath temp. °F*	*Current density, amp/sq ft*
1		16	2	2	5.0–5.5	room	5–10
2		9	0.7		5.6–6.0	90	10–20
3	6	32		4	4.5–5.6	115–120	20–100
4	6	32		4	5.6–6.0	150–160	20–100
5	6	44		5	1.5–4.5	115–140	25–100
6	40			4	2	140	25–100
7	23	26		5	1.5	115	100
8		24	3	4	5.6–5.9	110–140	25–50
9		20	4	4	5.0–5.5	75–90	

Table 14 Nickel Deposit Properties

Bath No.	*Hardness, VHN*	*Tensile strength, psi*	*Elongation, % in 2 inches*
5	140–160	51,000	30
6	230–260	98,900	21
8	425	152,000	6

cracking when heavy deposits are applied.

Bath No. 6 will produce deposits that are finer grained, smoother, harder and stronger than those of the Watts baths.

Bath No. 7 is a compromise between the properties of the deposits and the control problems of Bath No. 6 and the Watts baths.

Bath No. 8 is used for depositing hard, tough coatings, while Bath No. 9 is recommended for barrel plating.

Properties attainable from three of the baths are shown in Table 14 illustrating that appreciable differences in hardness, strength, and ductility are brought about by relatively simple variations of the baths.

Alterations of appreciable magnitude can be made by variation of the Watts bath formulation and operating conditions. An increase in pH, an increase in chloride content, or a decrease in temperature will increase grain refinement and hardness. Grain refinement is associated with occlusion of basic nickel compounds and can be induced similarly and more effectively by occlusion of organic addition agents in the deposit. Such refinement is usually associated with hardness, brittleness, high strength, and low ductility.

Other properties that can be favored by alterations of the bath include buffability, brightness, and machinability. When a specific property is important, a little experimentation will reveal how and to what extent it can be favored. Baths cannot, however, indiscriminately be changed without restriction or loss of the control range or loss of some other important characteristic. Further differences have been brought about by co-deposition of a second metal such as cobalt, and, of course, by deposition from other types of baths such as sulfamate and fluoborate.

Bath Limits

The number of baths that are published without limits is astonishing.

It gives one the impression that limits are not necessary or that a bath can be operated at a fixed concentration. Not only is this unture but it is difficult even to make a bath up to a fixed formulation. Some baths, like many of the lead baths, can successfully be operated to an approximate formulation and to wide limits; others, such as the conventional chromium bath, must be controlled to narrow limits. Sometimes broad bath limits are given to indicate that plating is possible over a wide range; however, it is not always clear whether the range represents a single bath or the possibilities of several baths within overlapping ranges. When nickel baths are recommended for quality plating and a fixed formula is given, it should only be regarded as a recommendation or as a starting point. In fact, a fixed formula, regarded only as a recommendation, often is a practical approach because of the many different applications of the nickel baths.

Bath No. 5 includes wide pH and temperature limits that may be narrowed for selectivity. The bath will deposit metal at consistently higher plating rates and with better throwing power when operated at 130 to 140°F rather than the wider allowable range. When operated at a narrower and low pH range of 1.5 to 2.5, anode corrosion will be favored. In a high pH range, corrosivity will be reduced. When anode size is limited by the work to be plated, anode corrosion can become an important aid to bath stability. If the work, the plating racks, or supporting equipment are attacked by the plating bath, then low corrosivity must be favored to reduce bath contamination. A wide variation of bath behavior and deposit properties is possible with the baths or modifications of the baths of Table 13. It also follows that a given bath will change appreciably with variations in concentrations, pH, temperature, and current density. Control limits may be shifted to favor a particular characteristic but operation will only be consistent after working limits are established through experience. If a change is made it should be from one set of limits to a new set of fixed limits. In actuality, a plating bath is not a plating bath until all important limits are set. This is particularly true of the versatile nickel baths.

Bright Plating

Practical and proven control limits applied to a No. 5 type bath are detailed in Table 15. This bath is operated within limits that are sufficiently broad for easy control yet narrow enough to assure consistent performance.

Primary control over this bath is executed by adjustments based on analysis for chloride and a gravity measurement. The chloride analysis indicates required additions of nickel chloride. The gravity is maintained by additions of nickel sulfate, thus eliminating the need for a sulfate analsis. Only occasional analyses for nickel are needed since the nickel limits are maintained by the total of the other controls. If the gravity is maintained and the lower nickel limit of 10 oz/gal cannot be maintained, then the entry of impurities into the bath is indicated. Occasional analyses for boric acid are essential since these limits must be held. The pH limits are quite narrow but are vital to optimum performance. The pH is raised by the addition of nickel carbonate or lowered by the addition of sulfuric acid.

In this bath, two proprietary brighteners were used that were controlled by plating tests. The gravity measurement was taken at 85°F rather than the more conventional 70°F because in the most concentrated condition salts crystallized at 70°F. Agitation was maintained by cathode rod movement.

With these complete defining conditions imposed on the bath, any failure to maintain set limits becomes significant. If bright plating canont be sustained at the prescribed conditions, corrective action is indicated. Occasionally, purification is required, and less frequently conditions are returned to normal by changes in the anodes or the racking. On rare occasions failure to meet all conditions will reveal an error in the overall control-operation-adjustment program.

Table 15 Bright Nickel Plating Bath

	g/l	*oz/gal*
$NiSO_4 \cdot 6H_2O$	260–340	35–45
$NiCl_2 \cdot 6H_2O$	45–67	6–9
H_3BO_3	30–45	4–6
Ni	75–83	10–11
pH	4.4–4.6	
Temp	115–125°F	
Agitation	10 ft/min	
Current density	40–50 amp/sq ft	
°Bé at 85°F	23.5–25.5	
Brighteners	as required	

General Purpose Nickel

Table 16 presents a useful variation of Bath No. 3. This bath will deposit good quality matte nickel. This is a useful general-purpose deposit that can serve as a protective or decorative matte nickel or as a beneficial undercoat in preparation for further plating.

Table 16 General Purpose Nickel

	g/l	*oz/gal*
$NiSO_4 \cdot 6H_2O$	220–260	30.0–35
$NiCl_2 \cdot 6H_2O$	30–60	4.0–8
H_3BO_3	20–40	2.7–5.3
pH	2.0–3.0	
Temp.	105–115°F	
Current density	20–30 amp/sq ft	

Activating and "Striking" Nickel Baths

The general-purpose bath or some variation of it is used as a strike to promote bond of a subsequent plate by the use of a nickel intermediate. Sometimes it will increase the reliability of the process by increased protection from the time the part is activated until it is overplated. Nickel serves a variety of preplate, activating, and striking functions that are not possible to separate one from the other merely because these functions are not well defined. However, the gain in reliability and the frequent necessity of the use of nickel as a process step is very clear. Low pH Watts baths can provide a striking function.

Nickel baths that are highly acidic and high in chloride are ready activaters. Several examples of these appear in Table 17. The room-temperature bath is used primarily for preplating of stainless steel. This is the most reliable general method for bonding an electroplate to stainless steel. It is accomplished by treating the steel anodically in the bath and then plating a bonding layer of nickel by cathodic deposition.

A bath that is a variation of the room temperature bath is the hot bath of Table 17. This bath is an analogue of the iron chloride bath, which is also a good activator. The advantages of this bath over the

Table 17 Activating Nickel Baths

Room Temperature Bath	*g/l*	*oz/gal*
$NiCl_2 \cdot 6H_2O$	220–260	30–35
HCl	30–38	4–5
Tem.	70°F	
Current density	20 amp/sq ft	

Hot Bath	*g/l*	*oz/gal*
$NiCl_2 \cdot 6H_2O$	400–440	53.0–59
HCl	7–10	1.0–1.3
Temp.	170–180°F	
Current density	60 amp/sq ft	

cold bath are that it can be operated at higher current densities and anode corrosion is excellent. It is not, however, commonly used for preplating of stainless steel. It is used as a bonding plate for steel following normal pickling. It is not recommended for parts that are stopped-off for selective plating.

Low Stress Nickel Baths

Nickel deposits with low stress and good ductility can be deposited from sulfamate baths. A basic bath, described by Barrett,[3] has been used extensively for electrotyping and electroforming. The basic bath contains

Nickel sulfamate	60 oz/gal
Boric acid	4 oz/gal

With proper control, low stress deposits can be produced at relatively high current densities. The use of addition agents, stress reducers, and pH control provide further dimensions to attain the desired properties.

The sulfamate bath can be modified by the addition of nickel chloride. This will aid bath conductivity and anode corrosion but it will also increase stress. Nevertheless, useful baths of this type are used that have some of the characteristics of the sulfamate bath and some of the advantages of the Watts bath.

Nickel Fluoborate Baths

Nickel fluoborate and nickel sulfamate have some common characteristics. Both of these salts are highly soluble and high concentrations of nickel favor plating at high current densities. The fluoborate baths will produce deposits with lower internal stress than the Watts baths, but they have the advantages of good conductivity and good anode corrosion and good bath stability that are more like the Watts baths than the sulfamate baths. These baths easily bridge the gap between the Watts baths and the sulfamate baths.

Bath Preparation

Bath preparation of table 13 type baths is carried out in rubber or plastic-lined tanks commonly used for nickel plating. The nickel salts are readily soluble in water, but the boric acid is sometimes troublesome to get into solution. The use of heat and agitation will greatly promote the dissolution of the chemicals. After dissolving the chemicals, the pH is adjusted and the bath is ready for initial plating. It is not likely that the bath will immediately plate satisfactorily. Very often, the deposit will be pitted or off color. If the deposit is satisfactory, as determined by a test panel, then it is often found that bond to the basis metal cannot be obtained. This condition is remedied by working the bath. Electrolysis will remove objectionable impurities and put the bath in operating condition. Electrolysis prior to plating is most effective at low current densities (3 to 7 amperes per square foot). During initial electrolysis the condition of the deposit will indicate the need for further bath treatment. If the deposit is pitted, a small amount of hydrogen peroxide will usually remedy the trouble. If the deposit is rough or off color, the bath should be treated with active carbon and filtered. It is good practice to work a fresh bath, treat it with active carbon, and filter it all at the same time. Small-scale tests can be made to predetermine the amount of treatment required and plating tests can be made on a large bath during treatment to follow the course of the treatment.

Operation and Control

The operating conditions of the nickel bath should be such as to obtain bath balance as closely as possible, with frequent adjustments to hold the bath within the range. If the bath is operated in approximate balance and adjustments are made to keep it in the range, then additions of chemicals will be small. This is important because excessive bath treatment will be avoided. The nickel salts contain impurities that make it necessary to treat a nickel bath before use. If large amounts of nickel salts are added during control, the same treatment will be required. However, small additions of nickel salts will not contain enough impurities to make purification treatment necessary. Small amounts of impurities will be eliminated during normal plating operations, without causing harm.

Usually, a nickel bath is operated according to some accepted formula. The limits on all ingredients will be set as follows: total nickel, sulfate, chloride, ammonia, boric acid, pH. All of these can be controlled by analysis and none of them should change rapidly in the balanced bath, with the exception of the pH. If the bath is operating correctly, the pH should increase and it should frequently be brought to the lower limit by the addition of acid. If initial data are taken on the bath, a graph can be prepared for additions of acid necessary to adjust to the lower pH limit for any pH reading (Fig. 14).

It will also be convenient to take gravity readings on the bath since a gravity reading is a rapid check on the over-all solids content. A second operating graph can then be prepared (Fig 15).

Bath balance is indicated if changes in composition are small. Changes in gravity, nickel content, or rapid changes in pH indicate that the balance is not proper. The balance is obtained by adjustment of the anode area. The cathode efficiency will be fixed by the bath

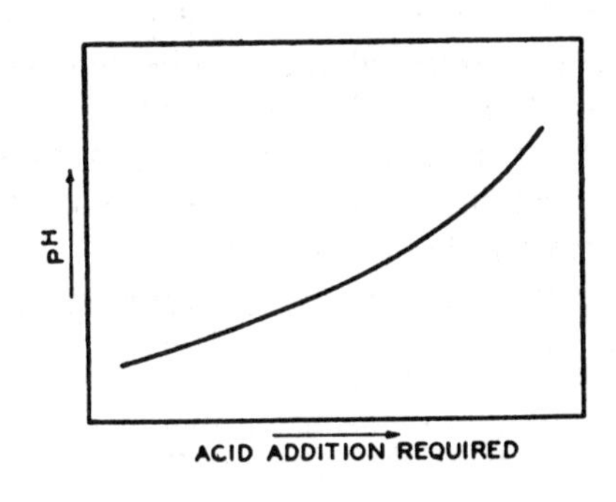

Fig. 14 Nickel pH vs. acid addition required.

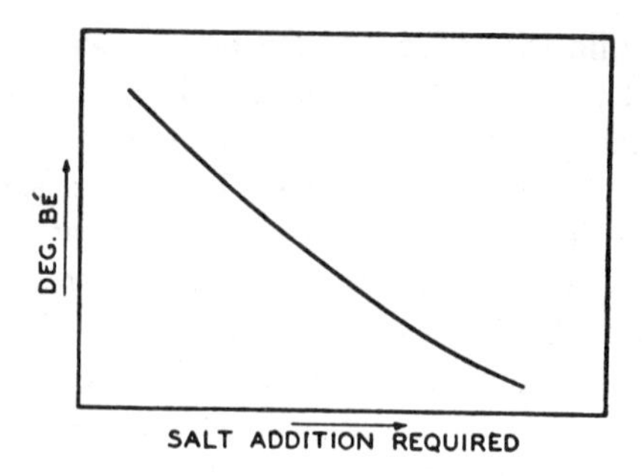

Fig. 15 Baume gravity vs. salt addition required.

limits; therefore, it will be necessary to adjust the anode area so that the anode efficiency is equal to or slightly greater than the cathode efficiency. If the positioning of the anodes limits the anode area (and thus the anode current density) then it may be necessary to try changing the cathode current density. However, if the cathode current density is dropped and the plating rate is not sufficient, then it is better to add more chemicals than to try to approach bath balance too closely. The working principle is to obtain a satisfactory deposit at a satisfactory plating rate and as close to bath balance as possible. Perfect balance, of course, cannot be obtained because drag-out will upset any scheme for perfection.

Anodes that will corrode uniformly, at the proper rate and without formation of sludge, are not available. But, of course, anodes are available that are satisfactory if they are accepted as not being perfect. Usually the corrosion rate is satisfactory and is accompanied by the formation of an insoluble deposit on the surface of the anodes. This deposit may become suspended in the bath and migrate to the cathode to cause a rough deposit and necessitate filtering. The trouble is overcome by bagging the anodes, but it must be realized that bags restrict replenishment of solution to the anode. If sludge builds up in the bags, the anodes will be further restricted so that they must be removed, cleaned, and replaced in clean bags occasionally.

Since nickel plates only slightly more readily than hydrogen (as indicated by a cathode efficiency of less than 100%), the nickel bath is subject to contamination by all the metals that plate from an acid solution, including iron and zinc. Usually the contaminating metals are removed by low current density electrolysis. Iron, which is a common contaminant, is removed by oxidation of the bath and increase of the

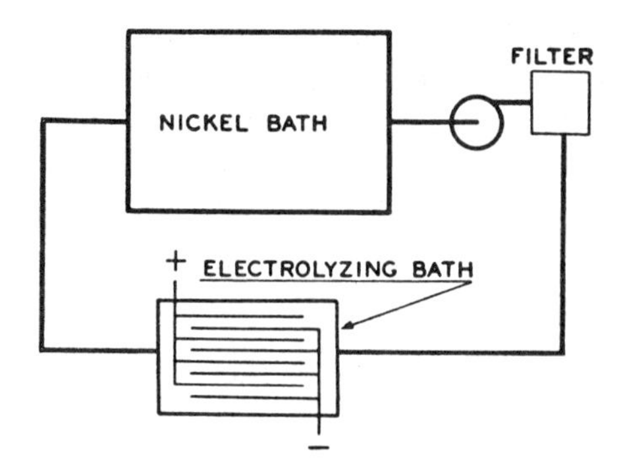

Fig. 16 Nickel electrolyzing bath.

pH with nickel carbonate until the iron is precipitated as ferric hydroxide. The ferric hydroxide is then filtered off.

Organic contaminants are common. These come from the chemicals, tank linings, stop-off lacquers, and rack materials. They can cause brittle and cracked deposits and must be removed. Electrolysis may remove organic contaminants, but active carbon is usually more effective. On large installations, it is profitable to filter, treat, and electrolyze a nickel bath continuously. This is best done in a tank separate from the plating tank (Fig 16).

Preparation of the Basis Metal

Steel is prepared for nickel plating by cleaning and pickling prior to plating. Pickling may be done by immersion or by the use of anodic current. In either case, sufficient acid should be present to remove rust or scale if it is present on the steel. The use of inhibitors in the pickling bath is not recommended. Inhibitors are used in pickling steel for economic reasons. The economy consists in saving acid by reducing the attack on the steel while scale is being removed. If steel is removed, of course, acid is used. Heavy scale is hardly ever encountered on work to be plated. In addition, it will be found that a certain amount of metal must be removed to render the steel active and to secure bond. It is best to remove this quantity of steel by pickling in uninhibited acid or by the faster anodic pickling treatment. In addition, inhibitors may cause trouble in the plating operation if they are not completely removed by the rinse.

If nickel is to be plated on zinc or aluminum, it is best to preplate these basis metals with copper before the nickel coating is applied. These metals are too active to be placed directly in an acid nickel bath. In the case of aluminum, it has been found necessary to either anodize or to immersion-plate zinc from a zincate solution prior to striking with cyanide copper.

Duplex Nickel

Remarkable advances have been made in the corrosion resistance of nickel-chromium plates through the use of duplex plating. This advantage is accomplished by depositing bright nickel over dull or semi-bright nickel and then depositing chromium. A tentative ASTM specification[4] has been proposed to aid the advance of this type of plating. Differences in potential between the bright and dull nickel have been observed—the bright nickel being less noble, thus tending to be slightly sacrificial to the dull nickel. Triplex nickel deposits also look promising and even further gains are made by the deposition of duplex chromium over the duplex nickel.[5] Proprietary processes are available for duplex and triplex nickel and for duplex chromium. In a broader sense these processes may be regarded as a type of layer-plating that no doubt holds much in store yet to be uncovered.

REFERENCES

1. W. L. Pinner, B. B. Knapp, and M. B. Diggin, *Modern Electroplating* (2nd ed.) The Electrochemical Soc. 1963 p. 260.
2. W. A. Wesley and W. H. Prine, *Practical Nickel Plating*, The International Nickel Company.
3. R. C. Barrett, *Proc. Amer. Electroplaters' Soc.*, **41**, 169 (1954).
4. ASTM B 375-63T, *Electrodeposited Coatings of Multi-Layer Nickel Plus Chromium on Steel.*
5. E. J. Seyb, *Proc. Amer. Electroplaters' Soc.*, **50**, 175 (1963).

22. ELECTROLESS NICKEL

The process that is commonly known as "electroless" nickel plating is a chemical reduction process that takes place on the surface of the metal immersed in the solution. The metal that is immersed must catalyze the reaction or else reduction must be started by another means. After the reaction is started it continues because the metal that is deposited continues to act as a catalyst. When the catalytic surface is removed the reaction stops.

Electroless nickel properly belongs in the plating shop as a process that competes with electrolytic nickel because of specific advantages over the common plating process. The major advantage is that the plating is only limited by the chemical reaction at the surface and not by the geometry of the bath as are electroplates. Thus, a uniform coating is deposited on all surfaces. The process is applied to complex shapes and intricate and small parts, such as gears that cannot economically be plated with the popular nickel plating baths. Valves, engine parts, tubes, hydraulic parts, electronic equipment, and fasteners are commercially plated.

Nickel deposits that are produced by chemical reduction are generally applied as an alternate to electrolytic nickel because the properties are quite similar. Thus, the major asset is corrosion protection of the substrate. However, chemical nickel is a nickel-phosphorous alloy containing 3 to 10% phosphorous. The properties therefore are different than nickel and vary with the phosphorous content. The deposits are semi-bright to bright and quite hard. Electroless nickel has good resistance to wear and this can be increased by heating to increase the hardness. The hardness can be increased from about 500 to 1000 Brinell by heating for one hour at 750°F. Heat treating produces a deposit that is competitive with hard chromium in wear resistance.

Electroless nickel will improve the solderability and brazability of steel and stainless steel. Also the hard low-friction surface will reduce the tendency of steel to gall, seize, and fret in contact with another steel surface.

The electroless nickel process was discovered at the National Bureau of Standards by Brenner and Riddell[1] and was patented[2] in the interest of the public. Since that time a great many patents have been issued[3] on control, applications and variations of the bath.

Bath Compositions

The basic chemicals to make up an electroless nickel bath are a simple nickel salt, a hypophosphite, and a salt of an organic acid. The nickel salts supplies metal, the hypophosphite provides reduction, and the third salt acts as a buffer and a complexing agent for the nickel. Many modifications of the original baths have been put to use. However, satisfactory deposits are obtained from the Brenner and Riddell baths. Typical baths comprise:

Nickel chloride	30 g/l
Sodium hypophosphite	10 g/l
pH	4 to 6
Temp.	194°F
Complexing agent	10 to 50 g/l

The complexing agent can be sodium citrate at 10g/l or sodium hydroxyacetate at 10 or 50 g/l.

These baths will deposit semi-bright coatings at a rate of 0.2 to 0.6 mils per hour.

Operation

It is a relatively easy matter to make up and operate a small bath. One should therefore not hesitate to do so if a potential problem exists that might be solved by the use of this coating means. If the limitations of the method are appreciated, good deposits should be expected.

It is recommended that small baths be made up with distilled water and reagent-grade chemicals. The solution should be filtered and kept clean. Glass, some plastics, and 300 series stainless steel may be used to contain the solution. The bath is then heated to temperature and the prepared part immersed until the desired deposit is attained.

During deposition the nickel and the hypophosphite are used and the pH changes. All of these changes decrease the rate of deposition so that the bath soon becomes inoperable unless plating is restricted to thin deposits, small parts, or unless additions are made during plating. With a little experience limited practical plating can be done by observing the rate of change of pH, by regularly checking the plating rate and by correlating these changes with the change in the plating capacity of the bath.

Increase in temperature markedly increases the plating rate. Overheating should be avoided and small hot surfaces should not be used to heat the bath. Local overheating can promote rapid bath decomposition.

Agitation is beneficial to avoid overheating and also increases the plating rate.

In general, preparation for electroless nickel plating is the same as for electrolytic plating, requiring cleaning and pickling, with the exception that plating will only start on a catalytic surface such as steel, nickel, or gold. It will also start on aluminum by being initiated by chemical displacement. On copper and silver, plating can be initiated by touching them with a catalytic surface. Plating also can be initiated by immersing the metal to be plated in a dilute, warm palladium chloride solution, rinsing and immersing in the bath.

Other Electroless Baths

Electroless cobalt and cobalt-nickel alloys can be deposited.

Proprietary baths for plating electroless copper and electroless gold are avaiable.

Metals can be plated by chemical displacement—such as copper plating of steel by immersion in a slightly acidified copper sulfate solution. This treatment produces a thin deposit that is limited in thickness by the fact that plating stops as soon as the substrate is completely covered. The deposit thickness is about one micro-inch (0.001 mil). Electroless nickel, by contrast, will continue to plate at a constant rate as long as the part is immersed and the plating conditions are held reasonably constant. Some baths will deposit metal much thicker than displacement baths but not as constantly as electroless nickel. One such bath that is used to plate tin on copper consists of stannous chloride and thiourea, acidified with either hydrochloric or sulfuric acid.[4] For example:

Stannous chloride	4 g/l
Thiourea	50 g/l
Sulfuric acid	20 g/l

This bath will deposit 8 to 18 micro-inches on copper in 5 to 30 minutes. The bath will continue to plate but at an ever decreasing rate, depositing 125 micro-inches in 24 hours. This is a simple, useful means to deposit thin coatings on copper for limited protection or solderability.

REFERENCES

1. A. Brenner and G. E. Riddell, *J. Research, Nat. Bureau of Standards*, **37**, 1 (1946).
2. U.S. Patent 2,532,283.
3. G. Gutzeit, *ASTM Special Technical Publication No.* 265, p. 53 (1959).
4. U.S. Patent 2,369,620.

23. PHOSPHATE COATINGS

Iron, aluminum, zinc, cadmium, and tin form insoluble phosphates or acid phosphates when these metals are brought in contact with acidic phosphate solutions.

It has long been known that a rust resistant coating will be formed when steel is treated with phosphoric acid. A simple immersion treatment is still used for phosphate derusting or preparation of steel for painting. Solutions of phosphoric acid and butyl cellosolve also are swabbed or brushed onto steel surfaces and then wiped off or washed off prior to painting. These treatments etch the surface and form a very thin phosphate film. They are better known as "metal conditioners."[1] More useful heavier "phosphate" films are produced by treatment in the well known *Parkerizing* and *Bonderizing* processes developed by the Parker Company.

Phosphate coatings are applied to provide a base for paint, oil, or wax or to reduce wear. The coatings are not recommended for outdoor exposure but they do have some protective value. They also have some appearance value and some of the treatments are used to produce "blued" finishes on steel.

Grey to black crystalline coatings are produced by a variety of proprietary acidic-metal phosphate solutions. The coatings are known as zinc phosphate, manganese phosphate, or iron phosphate.

Phosphate coatings are formed from the metal phosphates in the bath as well as by reaction of the acid phosphate with the substrate. When steel is treated with a zinc phosphate bath, a coating is produced that consists of insoluble zinc phosphate and some iron phosphate. The relative amounts of the two insoluble phosphates is controlled by the solution composition and the operating conditions. Excess acid promotes etching of the steel and reduces the coating efficiency. Excess iron in the bath causes troublesome sludge to form.

Zinc phosphate coatings of 100 to 1000 milligrams per square foot coating weight are used to promote paint bonding and improve the corrosion resistance of painted steel. Heavier coatings of 1000 to 3000 mg/sq ft are used to reduce galling of steel and as a base for lubricants.

The phosphate coatings are crystalline and the properties of the coatings vary with the crystallinity. Finely crystalline coatings that are more corrosion resistant are influenced by the surface finish prior to phosphating and are favored by operating at higher temperatures and by the use of addition agents. Coarsely crystalline deposits that can be favored by opposite conditions are desirable to retain lubricants on a phosphated surface.

Manganese phosphate baths form manganese-iron phosphates that are coarsely crystalline and more porous than the zinc phosphates. Coatings of 1000 to 4000 mg/sq ft can be produced.

Iron phosphate baths produce thin coatings of 30 to 90 mg/sq ft that are used for paint adhesion.

Phosphate coatings are applied by immersion or spray in a cycle generally as follows: (1) clean, (2) rinse, (3) phosphate, (4) rinse, (5) acidified rinse, (6) dry.

The crystalline character of the coating and the acid condition of the surface aid paint adhesion.

Deep drawing of steel parts is aided by the application of friction-reducing phosphate treatments.

Oiled and waxed phosphates are used to retard rusting of nuts, bolts, and small hardware.

Zinc and cadmium surfaces that are to be painted are treated with phosphates. These treatments are most economically applied as a supplement to the plating cycle: (1) plate (2) rinse, (3) phosphate, (4) rinse, (5) acid rinse, (6) dry.

Phosphate coatings are applied to aluminum surfaces to produce coatings that provide a good paint base and that aid deep drawing.

The phosphating process demands clean surfaces and good rinsing practices. Pickling is sometimes beneficial prior to phosphating and this is often done with phosphoric acid pickles. Sensitizing salts are sometimes added to the rinse preceding the phosphate bath to promote fine crystallinity. The acidified rinse after phosphating consists of a chromic acid or a chromic-phosphoric acid rinse sufficiently dilute to assure a slightly acid surface but at the same time avoid any attack of the coating.

REFERENCES

1. Military Specification MIL-M-10578, *Metal Conditioner and Rust Remover*.

24. SILVER PLATING

Silver plating, developed well over one hundred years ago, was one of the earliest electroplating successes: This process found a ready market in competition with rolled silver-clad copper "Sheffield plate." Electroplating proved to be a more versatile process than roll-plating, a process that could be applied to shapes as well as sheets, that could be applied to many metals, and could easily be applied as a thin layer.

Properties of Silver

Silver is a soft, white metal of pleasing appearance. It is malleable and ductile and an excellent conductor of heat and electricity. Silver is resistant to oxidation and corrosion but tarnishes readily in the presence of sulfide containing substances. The good conductivity of silver contributes to the use of the metal in decorative as well as technical applications.

A number of soluble silver salts are available and silver has been deposited from many different solutions of these. Silver provides a very accurate measure of a quantity of electricity when it is deposited from a silver nitrate solution under controlled conditions. Other solutions continue to be of interest but the silver cyanide bath, similar to the original bath, continues to be the primary silver plating bath.

Silver Cyanide Bath

Insoluble silver cyanide can be dissolved in potassium cyanide solution to form a soluble complex potassium silver cyanide:

$$AgCN + KCN = KAg(CN)_2$$

The double salt can be crystallized and is available as a convenient and readily soluble source of silver. Silver cyanide baths, however, are more commonly prepared from the slightly less convenient silver cyanide. In either instance the result is the same and a small concentration of silver ions is provided by ionization of the soluble complex.

Potassium cyanide in excess of the amount necessary to dissolve the silver cyanide is essential to the practical bath. The excess cyanide is known as "free" cyanide. Its presence extends the plating range, increases the limiting current density, increases conductivity, and aids anode corrosion. Carbonate or nitrate salts increase conductivity and broaden the plating range.

Addition agents are beneficial to produce bright deposits, semibright deposits, or merely to produce a fine-grained low-porosity deposit. Satisfactory deposits are being produced from a great many variations of the essential chemicals to make up a practical bath. It is really not possible to state that any particular bath is best or even that a specific bath is representative. There are, on the other hand, a number of baths that represent formulations that have gained popularity. One might conclude that almost any reasonable concentration of silver and free cyanide might make a useable bath. Economically, this is not true; the successful baths must be held within well defined limits of concentration, temperature, current, and cleanliness.

Baths made with potassium salts perform better than similar ones made with sodium salts. The potassium baths can be operated at higher current density, produce smoother deposits, have better conductivity, and respond better to addition agents. Sodium salts are cheaper than potassium salts but savings are insufficient to compensate for the loss of other plating values. Some savings can be realized without appreciable loss in productivity by the use of sodium salts and potassium salts.

Nitrates and hydroxides have been shown to be beneficial to some silver baths. In order to consider some of the variations of the silver bath, let us first consider a conventional silver-potassium-cyanide bath.

Basic Silver Cyanide Bath

	g/l	*oz/gal*
AgCN	40–50	5.3–6.7
Free KCN	35–45	4.7–6.0
K_2CO_3	40–100	5.3–13.3

This formulation represents a fundamental silver cyanide bath. It is a good general purpose bath and also can be considered as a "conventional" bath.

The basic bath is best made up from

	g/l
AgCN	48
KCN	65
K_2CO_3	40

The silver cyanide powder is added (with stirring) to a solution of potassium cyanide and potassium carbonate in one half the final bath volume. The KCN consists of 23 g/l to combine with the AgCN. leaving 42 g/l of free cyanide. It is good practice to make up the bath so that the silver and the free cyanide will be 2 g/l less than the recommended high limit for each. The potassium carbonate is made up to the low limit. It is advisable to use distilled water to make the bath. Tap water can be used but calcium carbonate will form due to the hardness of the water. It will then be necessary to allow the $CaCO_3$ to settle and filter the bath. Initial filtration can be avoided if the chemicals, the water, and the container are kept clean.

Potassium carbonate aids in the production of good deposits and adds to the conductivity of the bath. It is beneficial at concentrations of 40 g/l and more. It builds up in the bath with use and continues to be beneficial up to 80 g/l. Above this concentration the plating quality decreases with increase in carbonate but it only decreases markedly above 100 g/l.

This bath will plate satisfactorily at room temperature. The plating range can be extended by warming the bath and there is a continuing gain in plating characteristics as the bath is heated. Actually, the bath can be operated hot but heating greatly accelerates the decomposition of the cyanide and consequent formation of carbonate. At temperatures above 110°F losses offset gains.

The addition of small amounts of sulfide brighteners is useful. These will reduce porosity, crystallinity, and treeing. Carbon disulfide has been used for many years but it is critical to control. It should not be added directly to the bath. It is shaken with a portion of the bath or with a potassium cyanide solution and after standing one day a portion of the extract is added.

Sodium thiosulfate may be added directly to the bath. Additions at frequent intervals and at a rate of 0.05 g/l per day have proved to be beneficial.

Sodium-Potassium and Nitrates

	g/l
Ag	16–18
Free NaCN	15–22
Na_2CO_3	0–22
KNO_3	113–150

A bath containing nitrates and some sodium salts was described by Promisel and Wood.[1] The nitrate aids conductivity and promotes brightness. The bath will tolerate sodium salts, but at carbonate concentrations above 22 g/l maximum promotion of brightness is not attained.

This formulation demonstrates how a considerable amount of sodium may be included in the make-up and allowed to accumulate with maintenance additions. Also it shows that carbonate need not be added at make-up when another beneficial anion such as nitrate is used.

Nitrate-Hydroxide

	g/l
$AgNO_3$	40–50
Free KCN	30–35
KOH	2–5
Sodium thiosulfate	0.05 (daily)
pH	12.0–12.7
Temp.	90–100°F
Current density	6 amp/sq ft

A good quality silver bath can be made up with silver nitrate, cyanide, and hydroxide.

A bath of identical composition can be prepared from AgCN, KCN, KOH, and KNO_3. However there is an advantage to the use of silver nitrate. If all additions of silver are made as $AgNO_3$ then the nitrate is automatically controlled. The quantity of nitrate present is closely dependent on the silver additions and it happens that the resulting range of nitrate is beneficial. Silver nitrate is a cleaner salt than silver cyanide, due to the fact that it is a crystallized rather than a precipitated product. Silver nitrate is readily soluble in water and should be added to the bath as a solution. It should be added slowly with stirring to make sure that the initially precipitated silver cyanide redissolves.

Hydroxide in a silver bath will reduce cyanide decomposition. Carbonate will form at the same rate in such a bath but it will form at the expense of usage of hydroxide, for the most part, rather than cyanide.

The silver concentration is not critical in the warm nitrate-hydroxide bath. An increase in silver concentration will raise the critical current density and allow the use of higher current densities.

Engineering Plating

Heavy silver deposits are produced to provide stock for machining steel-backed, silver-lined sleeve bearings. A nitrate hydroxide bath was found useful to produce such stock[2]:

	g/l
AgCN	45–50
Free KCN	45–50
KOH	10–14
K_2CO_3	45–80
KNO_3	40–60
Temp.	102–113°F
Current density	75 amp/sq ft (agitated)

Many of the baths that are recommended for this purpose are much higher in silver concentration and in total salts. The limiting current density is increased by the higher metal content. However, to plate at high current density it is necessary to provide relatively rapid movement of the work with respect to the bath. The limiting current density is much more dependent on the movement of the bath, the work, or both than on the bath composition.

This bath is considered an engineering bath. If we regard it as a modification of the basic bath, then we might also suggest that such baths are engineering electrolytes by convention as much as by any marked difference in plating characteristics. Semi-bright deposits are obtained with ammonium thiosulfate as a brightener. The addition agent is added at 4-hour intervals at a total addition of 0.0015 g/l every 24 hours. It is not necessary that the deposits be bright but it has been found that the bright deposits have a far greater tendency to be free of pores.

Bright Silver Plating

Proprietary bright silver baths produce mirror-like deposits. These baths are similar to our basic composition with about double the free cyanide. The operating characteristics of the bath are superior to a nonbrightened type. Metal distribution, plating range, and throwing power are better. The deposit is harder and more resistant to staining. Most important, however, is the fact that processing costs are less when mechanical finishing can be reduced through bright plating.

Bright baths are widely accepted for the production of decorative work. The properties of the deposit are also of benefit to other applications such as electrical contacts. The bright baths usually require a primary brightener and a secondary agent such as a wetter. These are generally controlled by plating tests.

Operating Conditions

Silver baths are commonly operated at room temperature. This is generally satisfactory although the characteristics will change with temperature and large changes in room temperature will require some compensation in current or bath composition. As temperature increases it may be necessary to lower the current. This can be avoided by increasing the free cyanide in order to maintain the limiting current density. When this is done it means that the bath limits must be changed with substantial changes in temperature. It is perhaps better to select a bath temperature range that is above the maximum that will result from heating due to ambient conditions. A range of 95 to 100°F, for example, is easy to maintain with a small heater and overcomes the problem of introducing another variable.

A properly controlled quality silver bath will have an operating current density of 6 amperes per square foot or more. With cathode rod agitation the current density can be increased to 15. A cathode movement of 10 feet per minute will allow this. Plating rates can be incresed by stirring the bath but it is difficult to attain sufficient relative movement past all areas of the work by this means. If the shape of the work favors pumping of the solution over the surface it is then possible to deposit sound metal of almost any thickness at a current density of 75. At these conditions the plating rate will be 0.012 inch per hour. Higher plating rates are undoubtedly possible with efficient means of producing rapid relative movement.

Anodes are generally 999+ purity. Silver anodes are available as plates, balls, and cast shapes. The silver will dissolve at slightly over 100% efficiency due to complete conversion to soluble silver by electrochemical solution plus a slight chemical dissolution.

Particles formed or generated at the anode can cause rough deposits. These can originate from metal impurities in the anode that are not soluble in cyanide or from tiny silver crystals that sometimes dissolve loose from coarsely crystalline anodes. These particles form an anode sludge that can be retained at the anode surface by bagging. When bagging is impractical, anode troubles can be reduced by the use of 999.9 fine silver and by "worked" and annealed silver.

At high current densities or restricted solution flow across anodes, sometimes experienced by bagging, the anodes will form a dark, slowly soluble polarized film that can cause troubles very similar to those caused by impure or crystalline silver. Organic impurities in the bath may also contribute to this problem.

As a general rule the anode to cathode ratio should be 1 to 1. Solution movement that favors the cathode current density favors the anode equally well. Under rapid plating conditions, as used in engineering plating, higher anode current densities can be attained.

Preparation of the Basis Metal

When steel is plated with silver the steel must be activated and kept active until plating is started; by the use of a strike. The steel is cleaned of all grease and soil then activated by pickling in an acid. Following a rinse, the active steel is introduced into a silver cyanide strike—with the current on as the work contacts the strike bath. The work is "struck" with a thin silver deposit at low cathode efficiency, then transferred to a silver bath—again with the current on. This procedure assures a bond of the silver to the steel.

The thickness of silver required to strike steel is .006 to .012 mils. This thickness can be deposited in two minutes from the following strike:

	g/l
AgCN	1–1.5
NaCN	30–50

Common practice dictates that the strike be operated at 6 volts but it is good to know that results will be consistent if the current density is in the range of 50 to 100 amperes per square foot. If the silver concentration is doubled the strike time can be halved.

A single strike is reliable providing that the work is not held in the rinse following the pickle and providing that the work is moved readily into and out of the strike and into the silver bath. If holding time is required at the rinse following the pickle (for instance, to rerack the work), then a nickel plate should be applied prior to striking. The use of a nickel plate at this point is regarded as a nickel strike. It is not truly a strike in that it cannot promote bond without the aid of a true strike step to follow. A 5 minute deposit in a Watts-type nickel bath will suffice when this step is desired.

A low efficiency copper strike can be substituted for the silver strike. An advantage of the copper strike is that it will not promote rough silver plating due to over striking as will happen with the silver strike. However, results are no better than the silver strike with proper control.

Double striking is popular and triple striking is not uncommon. When these procedures are used each succeding strike is more concentrated in silver. Delays following striking and prior to silver plating are less critical with multiple strikes. A copper strike followed by a silver strike is not recommended. Such a practice is very liable to produce blistered plating. A good second strike is

	g/l
AgCN	3.5–5
NaCN	70.0–80
Current density	15.0–25 amp/sq ft

When two strikes are used the time should be halved in the first strike and reduced again in the second to accomodate the increased silver concentration or about 15 seconds for the above strike. A third strike can be a low concentration silver bath operated above the normal current density for a short period.

The second strike is used as a first strike to prepare copper alloy surfaces. A strike is also essential to plate over solder and zinc die castings. With the baser metals the striking action avoids immersion plating of silver on the basis metal—which avoids loss of bond. When immersion plating is a possibility a low silver concentration and a high cyanide concentration are likely to be successful.

There is no simple procedure that can be applied to all alloys. Brass, bronze, copper, phosphor-copper, and beryllium bronze all require different procedures. The same is true of the nickel alloys and the steels. In every case a pickle is needed that will leave the surface clean and active—then the strike is needed in addition. Unfortunately, a good appearance after pickling and after striking is not a guarantee of satisfactory preparation. The successful procedure is the one that works. It is usually easier to develope a cycle than to substantiate the tried-and-true. For this reason some people use three strikes and others use one.

Control

Broad control limits can be assigned to a bath that is employed for thin deposits. Chemical content, temperature, current density, and time may be liberally assigned. Narrower control limits must be set when heavier quality deposits are desired. Control of silver, cyanide, and carbonate is usually sufficient. Hydroxide or pH is controlled in some baths and nitrate, when used, in others.

Close chemical limits and control of temperature and agitation allow a high limiting current density. Such practice assures continuous plating under defined conditions.

Silver baths will tolerate a variety of common metals as impurities with no affect on the deposit. Sodium, copper, and iron, which are introduced by drag-in from strikes and by prolonged contact with steel, cause no trouble when present in relatively large amounts.

Excessive carbonate must be removed. It is removed by precipitation as calcium carbonate or barium carbonate. Calcium nitrate, barium hydroxide, or barium cyanide can be used for this purpose. These chemicals should only be added within the limits placed on cyanide or hydroxide or on the tolerance for nitrate.

A bath made up with cyanide and carbonate will have a pH of 11.2. An addition of 2 grams per liter of KOH will increase the pH to 12.0 and KOH can be easily controlled in the range of 2 to 5 by pH control of 12.0 to 12.7. If higher alkalinity is desired it should be maintained by titration control of the KOH.

The total concentration of chemicals in the bath is given approximately by the equation:

$$\text{Specific Gravity (at 77°F)} = 1.000 + 0.00057\,AgCN + 0.00061\,KCN + 0.00080\,K_2CO_3 + 0.00080\,KOH + 0.00050\,KNO_3$$

Specific gravity may serve to determine carbonate or nitrate when all other constituents have been established by analyses for each.

Plating tests are of value to control addition agents in bright baths. In other baths, a plating test is of occasional value to check the plating range.

Applications

Silver continues to have an appeal as a decorative coating on jewelry, musical instruments, and eating utensils. Even though silver is a soft metal and one that tarnishes readily, thin deposits in the range of 0.5 to 2.0 mils exhibit good life. The corrosion resistance of silver is sufficient to provide protection to specific chemical and laboratory ware.

Silver coatings of 0.1 to 1.0 mil are extensively used on electronic and electrical equipment. It also provides a base for gold where tarnishing must be avoided and for rhodium where resistance to wear is desired. Reflectors are silver plated and then lacquered to avoid tarnish.

High load, high-speed sleeve bearings are made with deposits of silver 2 to 20 mils thick and with an overlay of lead-tin deposit.

Silver applies where appearance, reflectivity, conductivity, solderability, or corrosion resistance are important. It is limited by diffusion when plated under or over copper, gold, or tin alloys and by its tendency to tarnish.

REFERENCES

1. N. E. Promisel and D. Wood, *Trans. Electrochem. Soc.*, **80,** 459 (1941).
2. R. A. Shaefer, *The Monthly Review*, American Electroplaters' Soc., **33,** 142 (1946).

25. ACID TIN PLATING

The application of tin to steel by electroplating was greatly increased during a period of tin shortage. Electroplating replaced hot dipping mainly because thinner coatings could be produced. Hot dipping methods for the production of tin can stock produced coatings of .07 to .09 mil and thinner controlled coatings could not be produced. Electrotinned coatings of .03 to .05 mils were easily produced. These thin coatings were flowed by heat, and where necessary the insides of cans were lacquered to aid corrosion resistance.

Satisfactory tin plate was produced both in acid and alkaline plating baths and ultimately 95% of all domestic tin plate was produced by electrolytic processing.

Characteristics

Tin has a pleasing white color and it resists corrosion and staining. It is a soft metal that does not compete with wear-resistant nickel coatings. Unfortunately, it is similar to nickel in that it is cathodic to steel. Therefore, it is essential to deposit pore free coatings in order to protect steel so that it does not compete with sacrificial zinc and cadmium. On the inside of a closed can, however, where the presence of oxygen is restricted, tin becomes sacrificial. Further, tin is non-toxic and thus widely used to contain foodstuffs as well as to coat food processing equipment.

Tin can be dissolved both in acids or strong alkalis to form salts that are sufficiently stable and sufficiently soluble to be adapted to use as acid or alkaline baths. The acid salts are bivalent (whereas the alkaline salts are quadrivalent), thus limiting them to a potential use of twice the quantity of current for deposition. Alkaline tin baths, on the other hand, present excellent throwing power so that there is no distinct advantage of one bath over the other.

Acid Tin Baths

The predominant necessity of acid tin baths is for addition agents. Without these, coatings from acid baths are loose, granular, powdery and treed. Consequently, the early history of acid tin plating was mainly a search for addition agents that would promote sound plating. Studies were made with many tin salts including tin sulfate. Quite a few agents were found that were effective, including cresylic acid, glue, beta-naphthol, resorcinol, cresol sulfonic acid, and others. Quite good results were attained by the use of at least two agents, and sometimes three and four were used.

Stannous Sulfate Bath

Pine[1] recommends the following specific bath:

	g/l	*oz/gal*
Stannous sulfate	54	7.2
Sulfuric acid	100	13.3
Cresol sulfonic acid	100	13.3
β-naphthol	1	0.13
Gelatin	2	0.27

The bath has a tendency to decompose. The stannous sulfate is gradually oxidized to stannic sulfate which has a low solubility; consequently, the bath continuously forms a precipitate.

The addition-agent problem and bath decomposition are definite disadvantages, but the bath has points in its favor. The anode and cathode efficiencies are both 100% and the bath can be operated at room temperature. High plating rates can be obtained and high current densities are made possible by the use of agitation.

Function of Ingredients

The stannous sulfate is a source of metal ions and the sulfuric acid increases the bath conductivity. The variation in concentration of these two chemicals can be very wide. If higher plating rates are desired, then a high tin concentration should be used. However, no higher tin concentration should be used than is required, since drag-out and bath decomposition will be increased by an increase in tin content.

Bath Preparation

The order of addition of the chemicals is not important, but some of the addition agents are difficult to dissolve. Glue or gelatin should be traeted with hot water to obtain a thick colloidal solution that may then be added to the bath. β-naphthol has a low solubility, but it may be dissolved by heating in a portion of the bath. It may also be added to the bath as a concentrated alcoholic solution.

The bath is highly corrosive so that plastic-lined or ceramic lined tanks are required.

Bath Operation and Control

The bath has to be largely controlled by plating tests or by carefully watching the deposit for signs of need for addition agent. The amount of control is somewhat dependent on the thickness of the deposit required. In any plating bath, satisfactory flash deposits can be obtained with less control than that required to produce sound heavy deposits.

The major ingredients can be controlled by chemical analysis, but for many of the addition agents, analytical methods are not available. For still plating, cathode-current densities of 10 to 40 amperes per square foot may be used. Anode-current densities should be kept below 25 amperes per square foot.

The sludge that forms in the bath does not present a problem since it readily settles. Occasional filtering may be required, however, if rough deposits are obtained. Filtering may remove a part of the colloidal addition agents, so that readjustment of the bath may be required after filtering.

Stannous Fluoborate Bath

The tin concentration and the acid concentration are not critical in acid tin baths. Limits may be chosen from economies imposed by the cost of maintaining a high tin concentration as compared to the gain in limiting current density at higher concentrations. When it is desired to operate at higher tin concentrations, a fluoborate bath shouble be preferred over a sulfate bath. This bath also offers the advantage that the stannic fluoborate is not limited in solubility to the same degree as the stannic sulfate. The same addition agents that are effective in the sulfate bath are quite likely to apply to the fluoborate bath.

Addition Agents

The addition agents used in the above sulfate bath play the following role in the fluoborate bath: The gelatine functions as a primary addition agent. β-naphthol will increase the limiting current density when used in the presence of gelatine. Cresol sulfonic acid will increase the limiting current density, decrease the tendency to tree in high current density areas, and function as an antioxidant for the stannous tin.

Bath Limits

The following bath has been used successfully:

Tin Fluoborate Bath

	g/l	*oz/gal*
Stannous tin	30–40	4.0–5.3
Free boric acid	20–30	2.7–4.0
Cresol-sulfonic acid	20–30	2.7–4.0
Gelatin (initial)	1	
β-naphthol (initial)	1	
pH	0.2–0.6	

This bath is not as well behaved as a Watts nickel bath but it can be operated continuously when control is supplemented by frequent plating tests. With the aid of such tests, deposits up to 2.0 mils were consistently produced at a current density of 20 amperes per square foot. One of the general virtues of the fluoborate baths is the high solubility of the metal salts, and the limiting current density can be increased in this bath by increasing the tin concentration. The tin can be increased to 60 grams per liter as stannous tin.

The bath is not critical with respect to tin, fluoboric acid, or boric acid. An acidity of 30 to 40 grams per liter will result in the recommended pH range, which can easily be controlled by pH papers. If desired, the total acidity can be controlled by titration of a concentrated sample to the first appearance of a precipitate. The boric acid does not respond to the normal titration but it can be held to satisfactory limits by estimating the concentration from the amount of boric acid it takes to saturate a sample (as described under Analytical Methods, p.277).

Current Density

Within the recommended current density limit the deposits from this bath are semi-bright. At higher current densities they are grey but still sound. The bath is typical of an acid bath and the limiting current density can be increased by heat or agitation. Heat will, however, increase the rate of formation of stannic tin, which does not contribute to plating performance, and it will accelerate the decomposition of the addition agents.

Other Acid Tin Baths

Sound tin can be deposited from fluosilicic acid tin baths. Such baths have long been used for electrorefining of tin, and there is a continuing interest in them because of a cost advantage of fluosilicic acid over fluoboric. These baths have little appeal to the electroplater because the quality of the deposits does not compare with the other acid tin baths.

A bath known as the du Pont Halogen Tin Bath has been used many years on a large scale for continuous electrotinning. The bath contains fluoride and chloride and the tin is stabilized as sodium fluostannite. Variations of the fluoborate bath are also available as commercially successful proprietary baths.

Anodes

High purity tin anodes are recommended to avoid noble metal anode sludges. Copper and antimony that are present in the anodes or are introduced into the bath will form such a sludge. Small amounts of these metals as well as lead will not interfere with the quality of the deposit but they are undesirable if the tin is to be used in food applications. On long use, insoluble tin salts can form on the anodes and then become suspended in the bath.

Preparation for Plating

Steel or copper are usually plated by conventional cycles consiting of an alkaline cleaning step and a mild acid etch prior to plating. A strike is sometimes necessary prior to plating cast iron.

REFERENCE

1. P. R. Pine, *Trans. Electrochem. Soc.*, **80,** 636 (1941).

26. ALKALINE TIN PLATING

Deposits from alkaline tin baths are essentially of the same quality as those from the acid baths; therefore, either bath may be used to produce electrotin plate or for many of the other common uses for food equipment or to produce a readily solderable surface. At times, though, the choice is dictated by the shape of the work or the way in which it must be plated. This is because the baths are quite opposite in many respects. The alkaline bath must be operated hot, it has excellent throwing power, cleaning prior to plating is much less critical than the acid bath, and no addition agents are required.

A bath may be made up either with sodium or potassium salts. The sodium salts are less expensive but the potassium salts are more soluble and more stable so that advantages accrue that often make it more economical to operate a potassium bath.

Sodium Stannate Bath

The following bath is recommended:

	g/l	*oz/gal*
Sodium stannate	90.0–120	12.0–16
Sodium hydroxide	7.5–11.2	1.0–1.5
Sodium acetate	10.0–15	1.3–2
Current density	20–40 amp/sq ft	
Temp.	165–175°F	
Anode current density	5–25 amp/sq ft	

The limits of the sodium stannate bath are almost self-defined. The plating rate is low compared to that of the acid bath. The plating rate can be increased by increasing the total tin concentration, but the solubility of sodium stannate is limited with build-up of carbonate in the bath. The following facts define the formulation and operation of the bath:

1. The cathode efficiency drops very rapidly with decrease in temperature.
2. The cathode efficiency drops with increase in caustic content.
3. The cathode efficiency drops with increase in current density.
4. Bath decomposition is increased with increase in temperature.
5. Bath stability is increased with increase in caustic content.
6. If sodium stannite is formed in the bath, it will decompose very rapidly and very rough or spongy deposits will result.
7. Sodium stannite is formed if the anodes are not properly polarized (white).
8. The anodes must be properly polarized (greenish-yellow to light-brown or gold in color) to obtain satisfactory bath performance.
9. If the anodes are excessively polarized (black), they will not be soluble.
10. The anodes will depolarize at a low anode-current density.
11. The anodes will become excessively polarized at a high anode-current density.
12. Continuous anode polarization is essential.
13. The anodes polarize more readily with increase in caustic content.

There are certain advantages that can be gained by changing any of the bath variables in either direction, but with each change, disadvantages also occur. The problem is really one of the magnitude of these changes and the operation must be governed by the known behavior of the bath. First, the bath must be operated at an economical plating rate, and in order to operate continuously, the anode area must be adjusted to obtain the proper anode-current density.

The sodium stannate is a source of metal and the sodium hydroxide limits are self-defined. The sodium acetate acts as a buffer and aids in anode-polarization control. Sodium carbonate builds up in the bath with use and has only minor effects on the operation characteristics.

Bath Preparation

The bath is made up in a steel tank provided with heating coils. The chemicals are soluble in water and they may be added in any order. Sodium stannate should be added in small amounts, because it is heavy and will lie on the bottom of the tank and large amounts on the bottom of the tank are difficult to get into solution.

Operation and Control

After the bath is made up, it is heated to operating temperature. A good deposit should be obtained immediately and it is often well to start operation with steel anodes. The bath will operate well with steel anodes and after it is in operation the tin anodes may be inserted one by one. By examination of the anode color, it will be obvious if the anode area of tin anodes should be increased or decreased. Some steel anodes may be used together with many tin anodes because the presence of steel anodes will prevent excessive polarization of the tin anodes. However, unless enough tin anodes are used at the proper anode-current density, tin will not be supplied to the bath at a sufficient rate. Steel anodes cannot be used exclusively and continuously because caustic soda will build up in the bath as tin is deposited. However, steel anodes are far less troublesome (and more satisfactory) for occasional plating and intermittent operation.

The anode problem and the cathode problem should be considered in view of the total current. For a desired plating rate, a certain current desnity will be required. The cathode area will then define the total current. From this total current and from the required anode-current density, the number of anodes can be calculated. As much tin as possible should be taken into solution since tin is cheaper in the form of anodes than as sodium stannate.

Analyses for tin and sodium hydroxide should be made several times a week, and additions of chemicals should be made daily on the basis of chemical consumption. If this is done, the fluctuation in caustic content will be small and anode control will not be difficult for continuous operation.

If sodium stannite is produced by depolarized anodes, the bath will turn dark and rough or spongy deposits will be obtained. This is readily corrected by adding hydrogen peroxide to the bath. Temporary operation with steel anodes will be best during such a corrective period.

With steel anodes and hydrogen peroxide available, anode troubles are not serious. A knowledge of the bath behavior is all that is needed to operate the bath and as experience is gained, operation will become easier, since the experienced plater can tell what is happening in the bath from the number and color of the anodes and from the bath voltage.

Characteristics of the Bath

The major disadvantage of the sodium stannate bath is that high plating rates cannot be obtained. However, for most applications, a thin deposit is all that is required and a high plating rate is not essential. The problem of close anode control is a second disadvantage, but the alternative of operating an acid tin bath is sometimes unsatisfactory.

The major advantage of an alkaline tin bath is that covering and throwing power are excellent. The characteristic drop in cathode efficiency with increase in current density prevents build-up in high current density areas and at the same time promotes covering in the recesses.

Potassium Stannate Bath

Much higher plating rates are attained with a potassium stannate bath,[1] particularly at the higher tin concentrations that are possible with the more highly soluble potassium salts. The sodium bath is greatly limited by the solubility of the sodium salts and the decomposition of the bath as the temperature is increased. Because of these factors it is impractical to increase the tin concentration, the temperature or the current density beyond the normal range. At 40 amp/sq ft the above sodium bath will operate at a cathode efficiency of about 50%. The efficiency can be increased by lowering the current density, but this of course results in a lower plating rate. Compare this with the following potassium bath of only average composition:

	g/l	*oz/gal*
Potassium stannate	177–202	23.5–27.0
Potassium hydroxide	20–25	2.7–3.3
Potassium carbonate	15+	2+
Current density	40–50 amp/sq ft	
Temp.	155–165°F	

At a current density of 50 the following cathode efficiencies were attained with this bath:

Temp., °F	*Cathode Efficiency*
120	35
140	60
160	90

This bath can be operated at low temperatures as a low-efficiency strike bath with excellent throwing power, at intermediate temperatures for higher plating rates and good stability, or at high temperatures for high plating rates at high cathode efficiencies. Baths at much higher plating rates can be made up by doubling the tin concentration.

Mixed Bath

A sodium-potassium bath can be operated to the following approximate limits:

	g/l	*oz/gal*
Sodium stannate	90–120	12.0–16
Potassium hydroxide	11–15	1.5–2.0
Potassium carbonate	30+	4+

This bath has some of the characteristics of the potassium bath but it is not greatly different than the sodium bath. If it is desired to extend the sodium bath slighly beyond its limits, this can be done by adding caustic as KOH and by adding some potassium carbonate to introduce potassium salts.

Anodes

The successful tin plater is the man who fully appreciates the problems of anode control. To be economical the anodes must be maintained in the filmed condition, the anodes must supply tin to the bath, the anodes must not cause unacceptable plating, and the anodes must not compromise the plating rate. By observation and care, optimization of anode control becomes routine but only providing that control time is devoted to the problem. Anodes of pure tin polarize and "film" in the characteristic and recognizable manner that is essential to operation. Tin anodes alloyed with 1% aluminum can be operated over a wider anode current density range and at higher anode efficiencies,[2] thus easing the anode problem.

Operation With Steel Anodes

Alkaline tin baths operated on a small scale can frequently be operated more economically with insoluble steel anodes. The reason is that the dominating anode problem is practically eliminated. The bath, however, changes when insoluble anodes are used and it should be thought of as a variation of another type of bath—the alkaline tin bath.

When tin anodes are used, a bath can be controlled within limits because tin is introduced into the bath and caustic is used. In addition, most of the tin is added to the bath as tin from the anodes, in a cheaper form than tin from sodium stannate. When steel anodes are used, it becomes necessary to add 2 to 3 times as much sodium stannate to the bath. Also, the sodium hydroxide builds up in the bath and as it does so the plating rate drops. Because of these factors it is definitely more economical to process on a large scale with soluble anodes.

Now consider the advantages of insoluble anodes. There is no anode polarization problem. The anodes are merely placed in the bath and allowed to remain. There is no anode current density range problem. Even the walls of the steel tank may be used as the anode. There is no production of stannous tin because the continual release of active oxygen at the anode keeps the tin in the stannic form. There is no control problem with sodium hydroxide since it is not added but rather is generated as tin is deposited. There is no need to add sodium acetate to the bath since it only aids the tin anodes. In addition to eliminating these problems, an advantage of greater bath stability is gained. There is less decomposition of sodium stannate to form sludge when the bath is allowed to increase in alkalinity.

A careful study is required to determine the volume of production necessary to operate a bath more economically with steel than tin anodes. Such a study was made with a bath processing 1000 square feet per day. It was found that with tin anodes about half as much sodium stannate was required as additions but that the savings in cost of tin would not pay for the time required to care for the anodes. This would not be true on a larger scale. When production is heavy the problem of increasing sodium hydroxide content also introduces additional cost factors when steel anodes are used.

When operating on a small, modest or intermittent scale, the use of steel anodes is definitely economic. Bath control is simple. All that is required is to add sodium stannate to take care of the tin deposited plus the drag-out loss.

Analyses should be made for tin which should be held within defined limits. Analyses should be made for sodium hydroxide and a log kept to determine the concentration at which the sodium hydroxide comes to an equilibrium. The cathode efficiency or the plating rate should be measured to adjust the plating conditions as the sodium hydroxide changes and to establish the plating conditions at equilibrium.

The plating rate can be increased by increasing the bath temperature or by increasing the tin content. A new equilibrium concentration will result from a change in bath conditions. The rate of formation of caustic will increase if the plating rate is increased by increasing the tin concentration. The rate of decomposition of caustic will increase if the plating rate is increased by increasing the temperature. By keeping a log, the best conditions to favor plating rate and bath life can be determined from the tin concentration, temperature, current density, and work load.

This bath is limited to low cathode efficiency, but on the other hand the cleaning power and the throwing power are superior to the high efficiency baths.

REFERENCES

1. F. A. Lowenheim, *Trans. Electrochem. Soc.*, **84,** 195 (1943).
2. F. A. Lowenheim, *Trans. Electrochem. Soc.*, **96,** 214 (1949).

27. TIN-NICKEL

An alloy that is attractive in many respects can be deposited from a chloride-fluoride acid bath. The alloy is bright and remarkably resistant to corrosion and tarnishing. It is deposited at about 65% tin, 35% nickel essentially as the compound SnNi. The deposit has a pleasant, slightly pink cast and good resistance to wear. The hardness of the deposit can be attributed to the properties of the compound. Apparently, the compound in an acid solution is also more noble than the component metals, resulting in a marked tendency of the compound to deposit in preference to either tin or nickel. Due to this circumstance the composition changes only a few percent with changes in current density.

This process, developed by the Tin Research Institute[1] and later by M & T Chemicals Inc.,[2] was used in England for plating of gear trains, watch parts, valves, pumps, and flow control devices. Although the alloy is hard, it retains some of the ease of solderability of the tin alloys and so has been applied to printed circuitry.

As described by Lowenheim,[3] the solution composition, bath make-up, and control limits are as follows:

Solution Composition

	oz/gal
Stannous chloride anhydrous ($SnCl_2$)	6.5
Nickel chloride ($NiCl_2 \cdot 6H_2O$)	40.0
Ammonium bifluoride (NH_4HF_2)	7.5
Ammonium hydroxide (NH_4OH)	to pH 2.5

	Make-up	*Control Limits oz/gal*
Stannous tin	4.0	3.5–5.0
Nickel	9.8	8.0–11.0
Total fluorine	5.0	4.5–6.0
pH	2.0–2.5	

Anodes of tin and of nickel can be used simultaneously but the recommended practice is to use nickel anodes and add the tin as stannous chloride.

The bath should be maintained to the recommended limits but plating tests are helpful to indicate the quality of the plating.

The coating, like those of tin or nickel, is non-sacrificial to steel and must be pore-free to protect a ferrous substrate. The practice of depositing bronze or copper as an undercoat has been found beneficial to promote corrosion protection.

The bath is normally operated at 150°F and 10 to 30 amperes per square foot. The cathode efficiency is practically 100% and at 25 amp/sq ft a deposit of 0.0005 in. will result in 15 minutes.

REFERENCES

1. Tin Research Institute, "Tin-Nickel Alloy Plating" (1952).
2. M & T Chemicals Inc. "Plating Tin-Nickel Alloy" (1961).
3. F. A. Lowenheim, *Metal Finishing Guidebook* p. 318 (1965).

28. TIN-ZINC

The tin-zinc alloys were studied by Angles[1] and the deposits are claimed to be superior in corrosion protection to either tin or zinc alone. Tin is corrosion resistant, but if pores are present in the deposit, steel will rust at these points. Zinc, on the other hand, protects steel at the expense of self-corrosion so that the steel will not rust at pores in the deposit. Unfortunately, however, the corrosion rate of the zinc is not entirely satisfactory. By selecting the proper tin-zinc alloy, a deposit can be obtained that has the corrosion resistance of tin and yet steel will not rust at small pores in the deposit.

Experimental baths were made up containing 30 grams per liter of tin. These were operated at a current density of 15 amperes per square foot and at 70°C. The analyses of the deposits are shown in Table 17.

The table shows that a wide range of compositions can be obtained. For maximum corrosion resistance, a deposit containing 78% tin was recommended.

The chemistry of the bath is complicated in that zinc forms complexes with both caustic soda and cyanide whereas the tin combines with the caustic alone.

The bath responds to the same variables that characterize an alkaline tin bath. The cathode efficiency increases with increase in tin

Table 17 Chemical Composition of Tin-Zinc Deposits

Zinc	*Total NaCN, g/l*	*Free NaOH*	*% Sn*
1.0	15.0	2.5	92
1.8	22.5	3.6	86
2.5	24.5	4.3	78
3.6	26.0	4.8	70
8.0	41.0	5.0	53
12.0	53.0	12.5	28

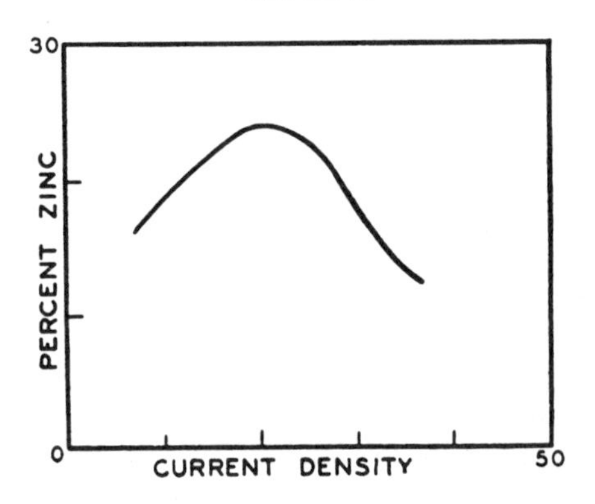

Fig. 17 Current density vs. zinc content of deposit.

content and temperature and decreases with increase in current density.

Alloy anodes are used, but they must be maintained in a polarized condition. Anode efficiency decreases with increase in anode current density.

An increase in caustic soda content decreases the cathode efficiency and increases the zinc content in the deposit, showing that the caustic soda concentration controls the tin-plating rate more than it controls the zinc-plating rate. Changes in the sodium cyanide content, above a concentration of 20 grams per liter, has very little effect on the composition of the deposit.

The change in the composition of the deposit with change in current density illustrates the complexity of the bath composition. The percentage of zinc is a maximum at a current density of about 20 amperes per square foot (Fig. 17).

The following bath is recommended to deposit 78% tin:

Bath 1

	g/l	*oz/gal*
Tin	30.0	4.0
Zinc	2.5	0.33
Sodium cyanide	25.0	3.30
Sodium hydroxide	4–6	0.5–0.8
Temp.	154–160°F	
Current density	10–40 amp/sq ft	
Anode-current density	7.5–15 amp/sq ft	

Properties

Alloys containing about 20% zinc are matte white with a tendency to stain. Outdoor exposure tests indicated that tin-zinc was superior to cadmium in industrial atmospheres and inferior in marine environments. The alloys have good solderability and have been applied to a number of electronic parts.

More Recent Baths

Work was done in England, subsequent to the work of Angles, on alkaline cyanide baths using sodium salts. Development work and applications of tin-zinc were done here with similar baths using potassium formulations.[2,3]

Bath 2

	g/l	*oz/gal*
Tin	38.0–53	5.00–7
Zinc	4.5–7.5	0.60–1.0
Free KCN	38.0–53	5.00–7
Free KOH	4.9–8.3	0.65–1.1
Temp.	145–153°F	
Current density	10–80 amp/sq ft	
Anodes	alloy	

Anodes must be kept polarized as they would in an alkaline tin bath to avoid generation of troublesome stannite in the bath. Preparation for plating is the same as for work to be alkaline tin plated.

The composition of the deposit is subject to change with changing bath conditions, typical of the alkaline-cyanide alloy baths. At higher temperatures the deposition of tin is increased. Increased free alkali favors the deposition of zinc whereas increased free cyanide favors the deposition of tin. Cathode efficiency is increased by increasing both metals and bath decomposition is decreased by decreasing the temperature.

Other Alloy Compositions

There has been some interest in other compositions, particularly 50/50 and 90 zinc, 10 tin, and it is obvious from the work done that a wide range of compositions can be deposited.

Tin-cadmium

If one is looking at the general potentialities of tin-zinc it is well to consider tin-cadmium at the same time. An alloy of 75 cadmium, 25 tin has been deposited from fluoborate baths for use on aircraft equipment subjected to marine atmospheres.[4]

Lowenheim has reported a potassium alkaline-cyanide bath that can be used to deposit tin-cadmium alloys.[5]

REFERENCES

1. R. M. Angles, *J. Electrodepositors' Tech. Soc.*, **21,** 45 (1946).
2. F. A. Lowenheim, *Modern Electroplating*, The Electrochemical Society, N.Y. (2nd ed.) p. 498 (1963).
3. U.S. Patent 2,675,347.
4. R. M. MacIntosh, *Encyclopedia of Engineering Materials and Processes,* Reinhold Publishing Corp. p. 675 (1963).
5. F. A. Lowenheim, op. cit. p. 503.

29. ACID ZINC BATHS

Zinc is popular as a coating for steel because it offers good corrosion resistance at a low coating cost. Since zinc is anodic to steel, it will protect the steel from corrosion even though the deposit is porous or contains small breaks.

Zinc may be applied to steel by galvanizing, sherardizing or metal spraying, but electroplating is preferable for certain requirements. Zinc coating by electroplating, sometimes called electrogalvanizing, has the advantage that the thickness of the deposit can be easily controlled. Also, the electrogalvanized coating is free from brittle iron-zinc compound layers that are formed in a hot process. Because the coating is not brittle, it may be applied to a sheet that has to withstand subsequent forming operations.

Acid zinc baths are used where it is desirable to have a high plating rate and low cost. The deposits are not as attractive as those from the bright cyanide baths and the throwing power of the acid bath does not compare with that of the cyanide bath.

The acid baths are primarily used for coating wire and steel strip.

Formulation

A typical formula for an acid zinc bath is as follows[1]:

	g/l	*oz/gal*
$ZnSO_4 \cdot 7H_2O$	360	48
NH_4Cl	30	4
$NaC_2H_3O_2 \cdot 3H_2O$	15	2
Glucose	**120**	**16**
Current density	10–30 amp/sq ft	
Temp.	75–85°F	
pH	3.5–4.5	

Zinc sulfate is used as a source of metal, ammonium chloride increases the conductivity, sodium acetate acts as a buffer, and glucose acts as an addition agent. Various formulations are possible and are equally effective as the one given. Other chlorides may be added to the bath, aluminum salts may be used in place of the ammonium salts or in place of the acetate, and also various addition agents may be used. The chemicals selected should supply metal content, act as a buffer, increase conductivity, and act as an addition agent. The same requirements hold for other acid baths, such as nickel, but the plating range of a nickel bath is readily changed by changing the formula. In an acid zinc bath the formula may be selected with more freedom, depending on the availability, cost, and purity of chemicals. Zinc sulfate is a high-purity commercial chemical made from a cheap commercial acid, sulfuric acid. The other chemicals in the formula given are produced on a large scale and are widely used in the chemical industries.

Bath Preparation

All of the chemicals used are readily soluble so that the order of addition is of little importance. After the chemicals are dissolved, the pH is adjusted and any required treatment of the bath is judged by the appearance of the deposit on initial plating. If the plate is rough, pitted, or off color, this may be due to suspended or soluble impurities. Suspended impurities can be removed by filtration or by allowing to settle. Soluble noble-metal impurities can be removed by low-current-density electrolysis, by immersion plating on the anodes, or by treatment of the bath with zinc dust.

Operation and Control

Although the bath is normally operated at current densities of 10 to 30 amperes per square foot, much higher current densities may be used if the bath is agitated. The current density is only restricted by the degree of agitation. For plating on wire, under special conditions, a current density as high as 2000 amperes per square foot may be used. If the bath is agitated during rapid plating, it is important that it be kept free of insoluble material; and therefore, continuous filtration is necessary. However, in still plating operations, occasional filtering or occasional removal of sludge from the bottom of the tank is sufficient.

The bath temperature should be maintained near room temperature so that if large currents are used, cooling becomes necessary.

The bath is not sensitive to changes in composition; therefore, chemical control is not a major problem. Control of pH is the most important factor. The pH gradually rises due to reaction of the acid in the bath with the anodes. When the pH rises to the higher limit, sulfuric acid is added to restore the lower pH limit. The acid reacts with the zinc anodes to form zinc sulfate at a rate that approximately balances the zinc lost by drag-out.

Anodes should be used[1] that are high purity and uniform in grain size. Zinc is an active metal and most anode impurities will collect on the anode as a sludge. If these are excessive, they will cause uneven corrosion and they may contaminate the bath with suspended impurities. In the latter case, the anodes should be bagged.

Characteristics of the Bath

The anode and cathode efficiencies of the bath are very high and the anode and cathode polarizations are low. Because of these conditions, bath balance and bath control are easy, but these conditions also cause the throwing power of the bath to be very poor. Thus, the bath is limited to plating simple shapes or to the employment of special racking or anode arrangements to obtain good metal distribution.

Preparation of the Basis Metal

Steel is prepared for zinc plating by cleaning and pickling. As in other baths, the cleaning procedure depends on the condition of the steel, but care should be taken that all grease is removed from the steel since the covering power of the acid zinc bath is poor and the bath has no cleaning power. The steel should then be pickled long enough to produce an etch but it should not be over-pickled. If difficulty is experienced in covering the steel, a strike in a cyanide zinc bath or in a cyanide copper bath will be helpful.

REFERENCES

1. W. Blum and G. B. Hogaboom, *Principles of Electroplating and Electroforming*, Mc-Graw-Hill, New York (1930) p. 316.

30. ZINC CYANIDE BATHS

Zinc cyanide baths are used to coat steel for protection from rusting. Thus, while they are used for the same purpose as the acid zinc baths, they have certain advantages over the acid zinc baths. First, the throwing power is good, so that irregular pieces can be easily covered. Second, bright deposits can be obtained, and bright deposits have sales appeal, even in applications where they are not required. Also, bright deposits do not stain as readily as dull deposits so that they remain attractive longer. However, it must be kept in mind that zinc is an active metal. This very property that is responsible for its good protection of steel is also a property that will cause the zinc to lose its original appearance much more rapidly than other electrodeposits.

Table 18 gives a direct comparison between the properties of the acid and cyanide baths.

Table 18 Comparison of Acid and Cyanide Zinc Baths

	Acid Bath	*Cyanide Bath*
Throwing power	Very low	Very high
Appearance	Dull gray	Semibright to bright
Basis metal	Can plate all ferrous metals	Cannot coat malleable and cast iron
Plating speed	Rapid plating is possible	Limited plating rate
Cost	Low	Low, but higher than acid baths
Tanks	Acid-resisting	Steel
Structure	Coarse-grained	Fine-grained
Control	Simple	Complex
Formula	Available	Proprietary for bright baths
Electrode efficiencies	High at all plating rates	High only under limited conditions
Elctrode polarization	Low	High
Preparation of basis metal	Clean and pickle but use of more care than for cyanide baths	Clean and pickle

Chemistry of the Zinc Cyanide Bath

A zinc cyanide bath is prepared from zinc cyanide or zinc oxide, sodium cyanide, and sodium hydroxide. Zinc cyanide combines with sodium cyanide to form sodium zinc cyanide:

$$2NaCN + Zn(CN)_2 = Na_2Zn(CN)_4$$

However, according to Blum and Hogaboom,[1] most of the zinc combines with sodium hydroxide to form sodium zincate:

$$2NaOH + Zn(CN)_2 = Na_2ZnO_2 + 2HCN$$

The chemistry of the zinc cyanide bath and many of the other cyanide baths is not as simple as represented by these equations. There is much known about the cyanide complexes and there is still much that is desired. For those interested in the complexity of these reactions, it is well to consult a paper written by M. R. Thompson.[2] For the operation and control of a plating bath, it is only necessary to know the chemicals used and the limits in which they must be held. For control purposes, we must know what the analytical method measures. For instance, in a silver cyanide bath, free cyanide is measured and controlled. In a cyanide zinc bath, the total cyanide is measured so that the analysis tells nothing about how the cyanide is combined. As previously pointed out, it is not necessary to know this. Yet it is convenient to know that the sodium zincate is present in a larger quantity than the sodium zinc cyanide, because one will then keep in mind that sodium hydroxide functions in a capacity other than to increase conductivity and pH.

Formulation

The basic zinc cyanide bath is described by Blum and Hogaboom.[1]

Plain Cyanide Bath

	g/l	*oz/gal*
Zinc cyanide	60	8
Sodium cyanide	23	3
Sodium hydroxide	53	7
Temp.	104–122°F	
Current density	10–20 amp/sq ft	
Cathode efficiency	90–95%	

This plain zinc cyanide bath will produce good deposits for protection of steel from rusting. If the bath is modified by increasing the total concentration and by increasing the temperature, the plating rate can be increased.

If a mercury salt is added to a zinc cyanide bath, an alloy deposit will be obtained that has a more pleasing appearance than a deposit from the plain zinc cyanide bath.[3]

Zinc Mercury Bath

	g/l	*oz/gal*
Zinc cyanide	37.5	5
Sodium cyanide	22.5	3
Sodium hydroxide	30.0	4
Mercuric oxide	0.25	0.03
Temp.	86–122°F	
Current density	40 amp/sq ft	
Anodes	0.1–1% mercury	

The bath is controlled in a manner similar to that of the plain cyanide bath except that one more constituent has to be controlled—mercury. If the mercury becomes high in the deposit, spots will develop on aging. If the bath is allowed to stand idle, mercury will immersion-plate on the anodes and the bath will eventually become depleted of mercury.

The zinc-mercury bath has good throwing power and good covering power and the deposit protects steel in the same manner as a plain zinc deposit.

Bright zinc deposits can be obtained by increasing the total salt concentration, by keeping the bath free of impurities, and by the use of addition agents.[4]

Bright Zinc Bath

	g/l	*oz/gal*
Zinc cyanide	60–82	8.0–11
Sodium cyanide	19–64	2.5–8.5
Sodium hydroxide	75–112	10.0–15
Temp.	82–100°F	
Current density	10–50 amp/sq ft	
Anodes	High-purity zinc	

Function of Bath Ingredients

Zinc cyanide is used as a source of metal in the cyanide baths. It is rendered soluble and modified chemically by sodium hydroxide and

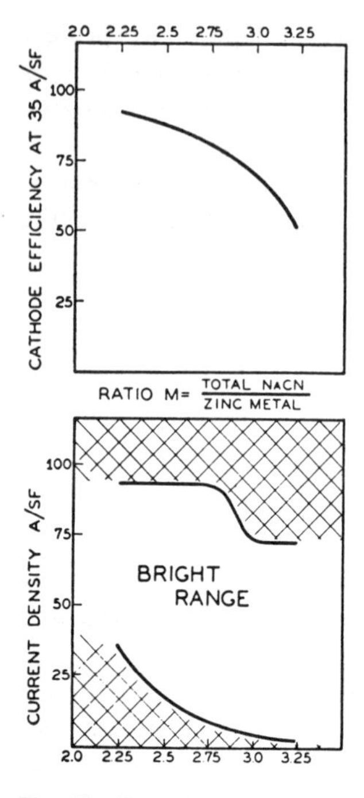

Fig. 18 Variation of current density and plating range with ratio.

sodium cyanide. These chemicals, in addition to forming complex zinc compounds, contribute to the bath conductivity. The ratio of sodium zincate to sodium zinc cyanide affects the plating-bath characteristics. The ratio of these complexes depends on the metal content and the hydroxide and cyanide content. The characteristics of the bath are controlled by holding the sodium hydroxide content within limits and controlling the ratio of sodium cyanide to zinc metal. This ratio ($R = \frac{\text{total NaCN}}{\text{Zn}}$) should be held within narrow limits to control the bath characteristics.

The ratio is held within the range of 2.2 to 2.6 for plain cyanide baths and of 2.0 to 3.0 for bright baths.

Hull and Wernlund[5] have shown how the bright plating range is shifted by the change in the ratio (R) and also how the cathode efficiency drops with increase in R. This data is shown graphically in Fig. 18.

It should be noted that the ratio is calculated from the total cyanide. This is the cyanide content obtained by analysis and includes the sodium cyanide plus the zinc cyanide calculated to the sodium cyanide equivalent. It is calculated from a fresh bath in the following manner:

Bath

	g/l
Zinc cyanide	60
Sodium cyanide	23

$$\text{Zinc metal} = \frac{60\text{Zn}}{\text{Zn(CN)}_2} = \frac{60 \times 65.4}{117.4} = 3.43$$

$$\text{Total cyanide as sodium cyanide} = 23 + 60\frac{2\text{NaCN}}{\text{Zn(CN)}_2}$$

$$= 23 + \frac{60 \times 98}{117.4} = 23 + 50.2 = 73.2$$

$$R = \frac{NaCN}{Zn} = \frac{73.2}{33.4} = 2.19$$

Carbonate will build up in zinc baths with use—due to absorption from the air and decomposition of chemicals. Usually, though, it does not become excessive since the build-up is rather slow and is compensated by drag-out. If cooling means are available, excessive carbonate may be removed by crystallization. The temperature at which it will crystallize depends on the total salt concentration and can be determined by cooling a small portion of the bath in the laboratory.

Bath Preparation

The bath is best prepared by dissolving the sodium hydroxide and sodium cyanide first. The zinc cyanide may then be added since it will be soluble in this solution. Other chemicals that are required may be added and the bath can be made up to volume. A plating test should then be made to determine the plating quality. The chemicals used for the zinc bath are of relatively high purity so that good deposits should be obtained immediately. If impurities are present, they will be revealed by an off-color deposit in the plating test. Or they will be revealed by a shift in the plating range in the bright bath which is sensitive to impurities. Noble-metal impurities can be removed by electrolysis or more readily by treatment with zinc dust followed by filtration.

Operation and Control

The zinc cyanide bath is controlled by maintaining the sodium hydroxide content within limits and by adjusting the ratio of zinc metal to sodium cyanide. The difficulty of maintaining these limits will depend on the type of bath used, the application, and the amount of work plated.

For higher plating rates and for barrel plating applications, higher chemical concentrations and higher temperatures may be used. However, the appearance of the plate will change with increase in temperature and if heavy currents are used, cooling will be necessary to keep the temperature from becoming too high.

The anode-current density should be held at 10 to 30 amperes per square foot to obtain proper corrosion. The anode-current density is obtained by adjusting the anode area, depending on the total tank current. Since zinc is chemically active in warm alkaline solutions, the anodes may go into solution faster than metal is removed at the cathode. This condition will be offset somewhat by the drag-out, but if it is excessive, it can be remedied by replacing some of the zinc anodes with steel anodes.

Bath balance depends on adjustment of the number and type of anodes so as to adjust the rate of solution of zinc to equal the rate of removal by plating plus drag-out. Sodium cyanide is then added to maintain the proper NaCN to Zn ratio.

Rapid plating conditions are the most difficult to maintain because the bath composition changes readily. However, such a bath can be maintained if frequent chemical and plating checks are made.

If the sodium cyanide content is high and zinc additions are required, zinc oxide may be added in place of zinc cyanide.

Plating to Limits

A zinc cyanide bath is better disciplined by plating to well defined narrow limits, as follows:

	oz/gal	*g/l*
Tl. NaCN	14.7–16.0	110–120
Tl. NaOH	13.4–14.7	100–110
Zn	5.3–6.0	40–45

These limits are recommended for general usage. If it is found that they are difficilt to maintain then some further change in procedure is indicated, such as a partial use of steel anodes or perhaps more frequent additions to compensate for drag-out. The narrow limits also encourage additions of zinc cyanide when zinc and cyanide are low and zinc oxide when zinc and caustic are low.

It is true that the bath can be operated more conveniently with broader limits and the ratio method of control but the performance is more consistent and less complicated with the narrow limit method. Also the benefits of any deliberate change to new limits are readily apparent when a change becomes desirable for more specialized plating such as barrel plating or continuous strip or automatic plating.

Preparation of the Basis Metal

The preparation of the basis metal is the same as for the acid zinc bath except that the cyanide bath has some detergent properties so that precleaning is not quite as critical.

REFERENCES

1. W. Blum and G. B. Hogaboom, *Principles of Electroplating and Electroforming*, McGraw-Hill, New York (1930).
2. M. R. Thompson, "Constitution and Properties of Cyanide Plating Baths," *Trans. Electrochem. Soc.*, **79,** 417 (1941).
3. U.S. Patents 901,758; 1,435,875; 1,451,543; 1,497,265.
4. R. O. Hull and C. J. Wernlund, *Trans. Electrochem. Soc.*, **80,** 416 (1941).
5. R. O. Hull and C. J. Wernlund, *Modern Electroplating*, New York (1942), p. 368.

31. CONTROL OF A PLATING BATH

A plating bath can be controlled by holding all chemical ingredients within specified limits, by plating-range tests and, in a few cases, with a hydrometer plus a simple test.

The control method selected depends primarily on the type of deposit desired. Baths used for flash deposits may be run for long periods without the aid of tests or analyses. For instance, an alkaline tin bath can be operated for months by merely using steel anodes and making frequent additions of sodium stannate based on the total current applied to the tank. However, if a high-effiency copper cyanide bath is used to produce 1 mil deposits the bath must be carefully controlled by biweekly chemical analyses.

It is then necessary to decide on control methods. The most frequently used and most reliable methods are those of analytical chemistry. If chemical limits are set on a bath and proved to be adequate, then the question of bath control is merely a matter of a few analyses. If the bath is analyzed and adjusted within chemical limits, attention can be paid to other plating variables such as temperature, racking, and the preparatory steps prior to plating. It is convenient to the plater to know that his bath is in the proper limits.

Good analytical methods are available but in many cases the choice of methods is rather limited. Methods that are rapid and of sufficient accuracy are not available to determine many impurities that enter a bath, and the impurities must be held below specified limits just as the chemical ingredients must be held within limits. If plating is being carried out on zinc-base die castings, zinc will build up in the bath since it is an active metal. When zinc is present in excessive quantity, it will cause trouble in most of the baths. Metal impurities regularly enter the baths by reaction with plating racks and the metal to be plated. In addition, metal is dragged in from the preparatory tanks.

Small amounts of lead will ruin the deposit from almost all of the alkaline or cyanide baths. Lead, tin, zinc, and chromium will cause trouble in a cyanide copper bath. Copper, zinc, chromium, and iron will contaminate a nickel bath, and any of the noble metals may cause trouble in an acid lead or acid tin bath. A complete list of metals that cause trouble would be difficult to compile because the sensitivity of the bath to the impurity changes with change in bath composition. The metals to suspect are those that are present either as metals or as metal salts in the plating room.

Since there are not enough practical analytical methods for the determination of metal impurities, it is obvious that chemical control cannot completely be relied upon. Also, it will often be found that a bath will not operate properly even though it is within the given chemical limits. This merely means that something is causing trouble that is not included in the analyses and that cannot be determined by the usual analytical methods. It may be the presence of an organic compound whose detection by analytical methods may be extremely difficult or no method may be available for its determination. Even the known organic substances that are added to the bath often cannot be controlled by analysis.

Although analytical methods are reliable for general plating-bath control, they do not guarantee good plating. Therefore, a plating test is needed. The simplest plating test consists of plating on a flat sheet of metal or on one of the articles regularly plated in the bath. However, such tests only detect trouble after it has occured; and the primary purpose of testing and analysis is to keep out of trouble.

If a plating-range test is employed, then trouble may be forestalled. This is because troubles often start at either high or low current densities and creep into the normal plating range. The plating-range test, such as the Hull test or the bent-cathode test, pick up troubles at current densities outside the normal plating range and corrective steps can be taken before trouble occurs at the plating tank.

There are very few baths that cannot be controlled by chemical analysis and a plating-range test. However, these tests are not required in every case and they are not necessarily adequate in every case. Let us assume that we have a fully automatic plating machine where the time and the current in the plating tank is limited and the thickness of the deposit is specified. It will be necessary to control the plating rate. Chemical analysis, temperature control, and current control do not guarantee the cathode efficiency. The cathode efficiency may drop as impurities enter the bath even though these impurities do not harm the quality of the deposit or are not revealed in a plating-range test.

The plating rate is measured either in terms of thickness of metal deposited in a given time or in terms of cathode efficiency at a given current density. If it is measured as thickness, it may be measured on a test panel; but more often, it is measured directly on one of the plated articles. Such measurements may be made by magnetic means with a proper combination of metals or by other mechanical, or instrumental means. The usual chemical methods consist of determination of the time necessary to dissolve the deposit, or stripping the deposit from a known area, followed by analysis for the total metal stripped. Cathode efficiency is measured by plating a weighed test piece for a known time and current and reweighing to determine the amount of metal deposited.

The selection of analytical methods depends on the speed and accuracy required as well as where the method is to be run. It is often convenient and sometimes necessary to carry out tests and analyses in the shop. Also, it is often true that some of the ingredients can only be determined in a properly equipped analytical laboratory.

Most analytical methods that have been adopted to plating control are relatively simple, as analytical methods go, and can be run with a minimum amount of chemical equipment. A few, however, such as the determination of Rochelle salt, are relatively complex. These more complex methods must be run with proper equipment and by a trained analyst. Many of the methods, however, can be run at or near the plating line. Then too, it will be found that the simple methods are most successful when they are run in the plant.

For rapid control methods, it is best to sacrifice some accuracy for the sake of simplicity. Furthermore, the closer the control method is to the line and the sooner the results are used for bath correction, the less need there is for a high degree of accuracy. The analyst who does not know what the tank is doing can only make required analyses and guarantee his accuracy from his knowledge of analytical chemistry. He may not know how his results are used to correct the bath. The plater, on the other hand, may run a plating-range test and suspect low metal content. If he can confirm his suspicion by some approximate rapid check on the metal content, he can add metal and run a second plating-range test. This may be done in less time than is required for sampling and transferring the sample and the results to and from the laboratory.

The plater, with a few simple tests, can often do a good control job, and it takes the mystery out of control work. If the plater has never seen a laboratory or does not understand the analytical procedures, he may not have sufficient confidence in the numbers he receives from the laboratory. Improper sampling can cause no end of trouble between the plant and the laboratory for the analyst can only analyze what he receives. In some cases, the plater may take a sample before his salts are completely in solution on making up a fresh bath. Many plating salts, such as boric acid, dissolve very slowly, and others, such as the alkalis, will form large slowly dissolving lumps at the bottom of the bath if they are not added properly. If the plater has a rapid control method at the line, he will learn what is happening when he makes up a new bath and when he makes additions.

A rapid over-all check can be made on any fresh bath with a hydrometer. The fresh bath must contain some total salt concentration made up from specified quanitities of various chemicals which will always be of the same specific gravity. The hydrometer is a valuable tool that should be used wherever possible because it is extremely rapid and highly accurate. After a bath has aged and accumulated impurities or chemical decomposition products, the hydrometer may be useless, but it will always apply to a fresh bath; in many cases it gives reliable information throughout the life of the bath. It is possible, for example, to control a chromic acid bath with nothing more than a hydrometer and a plating-range test.

A conductivity measurement is similar to a gravity measurement in that it is rapid and is affected by all the salts present, but it does not measure the gravity. A conductivity meter responds principally to the hydroxyl ion in alkaline solutions and principally to the hydrogen ion in acid solutions, for these are the ions that carry most of the current. It is applicable to direct control of electrolytic alkali cleaners because current-carrying capacity is of prime importance in such a cleaner and the hydroxyl ion usually does a great deal of the cleaning work.

The conductivity meter may be used in conjunction with some other test. If an acid and a salt are present, then both can be determined by measurement of conductivity and gravity. The acid primarily determines the conductivity, but a correction must be made for the salt present. The gravity measures the total solids present. Now by setting up proper original standards both the acid and the salt may be measured. This procedure works well where one substance is of high conductivity and the other of low conductivity. A temperature measurement must also be made, or all tests must be carried out at the same temperature, for the temperature has a considerable effect on both conductivity and gravity. Determination of the amount of acid and the amount of iron in a pickle solution is a typical example of this type of control.

It is possible to develop other rapid control methods, such as the measurement of cathode efficiency. In many of the alkaline baths, the cathode efficiency is low at high current density and the anode efficiency is high at low current density. Advantage may be taken of this by plating a cathode for a definite period of time at a high or at operating current density and then measuring the time for deplating the cathode at low current density. From the data obtained, the cathode efficiency at the operating current density can be readily calculated. In some baths, such as a silver strike, the cathode efficiency is a direct measure of the metal content and its measurement is thus a rapid method of determining the silver content of strike baths.

When unusual methods are used for solution control, the property that is being measured should be kept in mind. Thus conductivity may measure the alkali content of a cleaner solution, but in an electrolytic cleaner, the conductivity may be a more important control factor than any of the chemical substances the analyst may determine. If cathode efficiency is used as a means of checking metal content, it may be more important than an analysis since the plating rate is being measured and a direct control of plating rate may be more important than that of the chemical content.

One can go far with simple nonchemical methods, but the simplicity of many of the chemical methods should not be disregarded. As an example, direct determination of free cyanide in a silver bath is a simple, rapid, and accurate chemical method. It is doubtful if any other method is better than direct titration to determine free cyanide in a silver bath. Many other titrations, such as for free acid, are just as simple and quite easily run at the plating line in the same manner as in the laboratory. A few solutions and simple glassware are all that is required. However, many methods can only be run in a properly equipped laboratory.

Solution control is a must for successful plating. There is no plating bath that will adjust itself to the proper equilibrium by electrolysis. But solution control is not the whole answer. If too frequent additions are required for continuous quality plating, then the bath is too far from balance. A perfect balance cannot be maintained, but it should be approached. If the bath changes rapidly, an adjustment of anode and cathode areas or of current should be made until balance is approached. Large changes in bath composition should be avoided since such changes require extensive adjustments that may result in sudden changes in bath properties and a shift in the plating range. A large addition often requires some other secondary correction and results in loss of production time.

It is often helpful to keep a daily log of a bath, showing chemical composition, tests, additions, current density, production time, temperature, and appearance of the bath and of the anodes. It is also revealing to plot this data against time, and it will often be found that the bath will change over a period of weeks and that at times this can be coordinated with the number of anodes or some other data. Changes from day to day are difficult to see, but a gradual change over a period of months is apparent from a plot of the log data. Such a plot may reveal that additions become much larger and corrections more frequent when the total current is increased. This may be due to the fact that the tank is not sufficiently large. If a tank is crowded with work and run continuously, it will be more difficult to control than the same amount of work in a larger tank.

Overworking a tank may lead to rapid build-up of impurities. Impurities can enter the bath from everything that comes in contact with it—the racks, chemicals, the water, the air, and the articles to be plated. The impurities are of two general types: those that will plate and those that will not. The metals that plate more readily than the metal in the bath cause trouble. They cause the most trouble in low-current-density areas and that is where they will be revealed on a plating-range test. The plating-range test also gives the clue as to means of removal of the impurities. By electrolyzing the tank at low current density, troublesome noble metals are removed. Metals or other impurities that do not plate usually are not harmful unless they build up in large amounts. It will then be necessary to remove them by crystallization, by chemical means, or by dumping the bath.

Control of a plating bath is usually accomplished by standard analytical methods, however, control procedures for a specific application should be dictated by experience and experimentation.

32. PLATING TESTS

A plating bath will operate satisfactorily when all potential troubles are controlled. Obvious sources of trouble are avoided by routine maintenance of established chemical limits. Less evident are those introduced by contamination from chemicals, rinse water, anodes, tank linings, plating racks, the atmosphere, and decomposition of the bath.

Complete bath regulation includes primary control of the major constituents by chemical analyses, gravity, pH, and secondary control by plating tests.

Secondary control includes testing to determine the needs for addition agent adjustment, filtration, and purification. These needs are "read" from a plated test panel and corrective steps are deduced from sample scale additions and retesting.

A bath may be tested directly by plating a few pieces of work at higher than normal current or by plating complex shapes. This may reveal rough plating in high current areas and warn of the start of trouble that can spread later to the normal current areas. In a like manner the spread of trouble is indicated when it becomes necessary to drop the current below normal in order to maintain the quality of the work.

Troubles are better detected by plating a test panel so positioned that a range of current densities is produced across the panel. A plating range is produced by confining a panel nonuniformly within insulating walls. The well known Hull cell[1] is the most widely used of the plating tests. The best known configuration consists of a quadrangular box with sides of unequal lengths as shown in the plan view of Fig. 19. The test panel is inclined at an angle of 37° with respect to the long side of the box. The anode is perpendicular to the long side so that the anode-to-cathode distance changes in a regular manner along the cathode. This results in a continuous change of current along the panel covering a wide range of current densities.

Some prefer to use a bent cathode because it covers a wide range

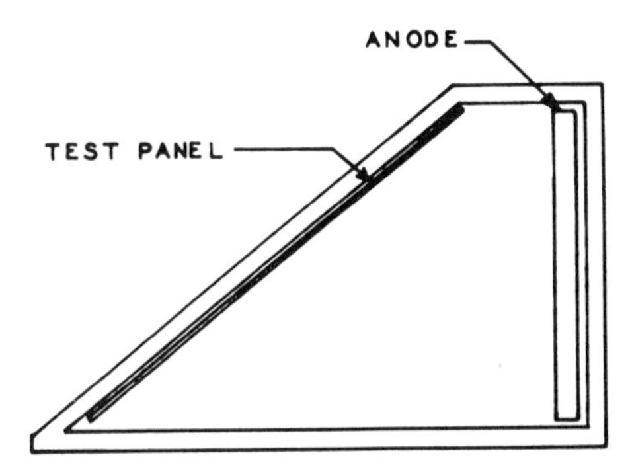

Fig. 19 Hull cell: plan view.

and it exposes a corner as well as vertical and upper and lower surfaces to the electrolytic action. Thus, it tests current range, throwing power, settling particles, and partially-trapped gas. It can be placed in many positions within an insulating box to favor particular areas on the cathode. To have meaning it must be reproducibly placed and it must be compared with standardized cathodes. One arrangement is shown in Fig. 20.

Some prefer to confine a cathode within a box, to keep the anode outside the box, and to plate through a slot. Since the current passes through the slot, it acts as a current regulator, and for the practical purposes of current distribution it serves as an anode. This scheme eliminates anode troubles that arise from placing an anode within a limited space inside a box.

The major trouble with an anode is polarization. Lesser troubles are anode sludge, reproducible placement of a corroding anode, and anode convection currents. These troubles all become insignificant when the anode is outside the box. Fig. 21 shows one arrangement of a slot-type cell.

This cell can be used as a dip cell directly in a plating tank. Modifications have been used—as an integral cell with an anode compart-

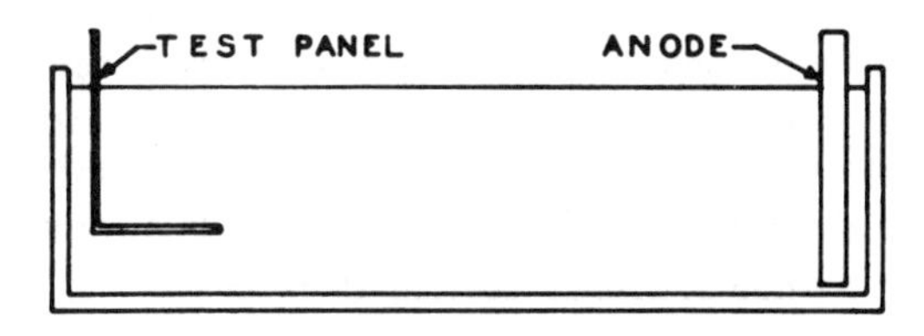

Fig. 20 Bent cathode: elevation.

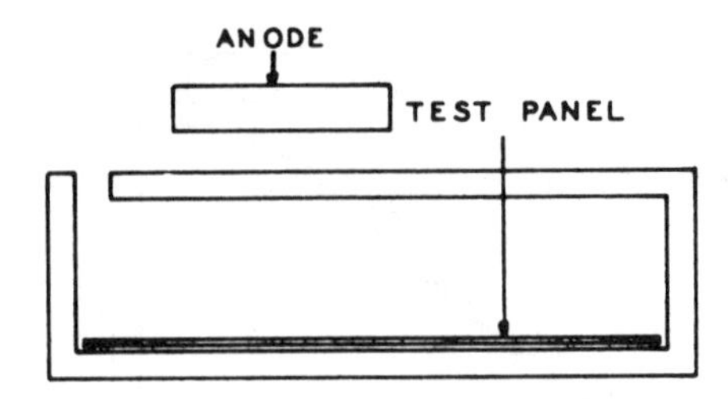

Fig. 21 Slot cell: plan view.

ment, as a vertical cell with the slot at the bottom rather than at an end, and as a mutiple slot cell to change the current distribution.

For maximum advantage a plating cell should be calibrated, so that the current range can then be read in terms of current density. However, for control purposes direct qualitative comparison is all that is required.

When a plating test is qualitatively used it is beneficial to have a test panel that was plated when the bath was under optimum control. This condition is generally achieved after the bath has been freshly prepared, plated for a short period of time and filtered. Any other initially required adjustments such as pH, surface tension, or gravity should be made.

A test panel should be plated at a standardized time, current, temperature, solution level, and type of agitation (if used).

It is advisable to record the appearance of a panel; this is done by a direct sketch of a panel section as shown in Fig. 22.

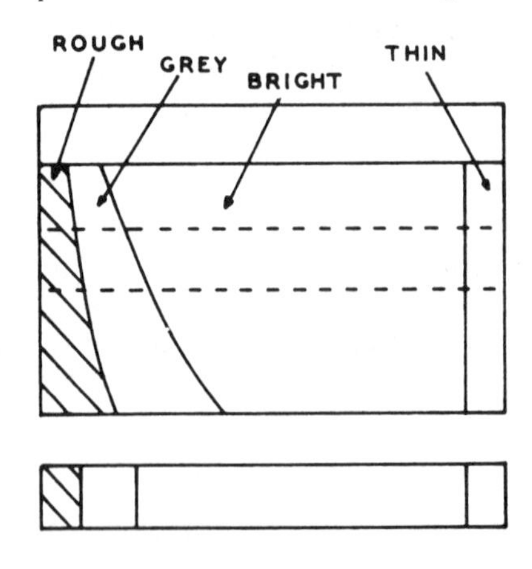

Fig. 22 Test panel (top); recorded section (bottom).

Routine Testing

Before sampling, an operating bath should be brought to level and thoroughly agitated. A sample portion should be transferred to a clean test cell: A cell that is dirty or is contaminated with another type of bath can give misleading results. The temperature of the test portion should be adjusted or controlled to the bath operating temperature.

Clean test panels of identical surface finish should be used for comparative testing. Commercial zinc-coated steel and polished brass panels are available. Zinc-coated panels are stripped in dilute hydrochloric acid to produce a clean and active surface just prior to use. Brass panels are used when it is desirable to have a background that contrasts with the deposit or to test on a copper alloy substrate.

A current range to 10 amperes (at up to 12 volts) an ammeter, and a rheostat provide the current and control needs to test 4 inch panels. Tests are usually done, after a standardized surface preparation, at 1 to 3 amperes for 1 to 10 minutes.

It is advantageous to test with anodes of the metal being deposited. This minimizes changes due to testing and up to 6 or 8 consecutive tests can be made with one sample in a small cell. Polarization of the anode, within a small cell, can be reduced by the use of a corrugated anode. When necessary, insoluble anodes may be used but only a few tests should then be made on a single bath sample.

During a test, agitation usually is not necessary, even when results are applied to an agitated bath. When it is desired to simulate agitation this is done by moving a rod just in front of the cathode or by air agitation.

The panel can be read if experience with the test has been accumulated. When the plating range is in the low current area of the panel it is likely that addition agents are needed. Rough, dark and irregular appearance in the high current areas might indicate metal impurities. Pitted plating calls for a reduction in surface tension. Cracked plating often means that excessive or decomposed addition agent is present. Poor coverage in the low current area can be due to the same cause. Excess carbonate in cyanide baths will cause a slight darkening or increase in crystallinity of the deposit.

Corrections in the bath are made to remedy the suspected deficiency —addition agent, for example, is added to increase a lowered plating range. A second test will then reveal the extent of recovery if any. A series of tests with small continuing additions will disclose the optimum effective addition. It will also indicate the amount that will exceed the optimum so that over-correction will be avoided.

Organic impurities, excess addition agent, or decomposed addition agent are tested with activated carbon followed by filtering and retesting. Fresh addition agent and a third test will complete the analysis.

Metal impurities are often removed by low current electrolysis. This is done on a small scale at 3 to 7 amp/sq ft at noted intervals of current in terms of ampere-minutes per liter. Periodic plating tests are made at the noted intervals. This will reveal the removal of impurities and also reveal the current-time requirement per volume to correct the situation.

It is now seen that a bath can be doctored by starting with a test, making changes on a sample scale, and retesting until diagnosis and cure can be affected. This procedure is in fact often essential when unexpected trouble strikes the plating line.

A plating test is best applied after a bath has been analyzed and adjusted to the control limits. Tests may indicate acidity, free cyanide, or metal content out of limits but these should be adjusted by primary control, which is then supplemented by secondary plating test control. Plating tests just like analyses are best applied to keep out of trouble rather than to get out of trouble.

REFERENCES

1. R. O. Hull, *Amer. Electroplaters' Soc.*, 1939, Proc. 27th Ann. Convention.

33. GRAVITY, CONDUCTIVITY, AND VOLTAGE

There are times when the simple approach is best, times when we should ask ourselves the question: What is the simple way to do the job? Rapid, easy, on-the-spot measurements are often vital to keep a process going.

A very easy, uncomplicated measurement is made by dropping a hydrometer directly into a solution. It is sufficiently rapid that time can profitably be spent making speculative gravity measurements. It is a worthy practice to make a record a gravity measurement every time an untried bath is made up for the first time. This plain figure establishes the gravity as a consequence of the total dissolved solids. It verifies that the concentration of salts were present to support the measurement and it may suggest a ready means of simplifying solution control.

Another simple measurement that can be made is conductivity. The remarks that have been made about gravity can also be made about conductivity. It is well worthwhile to consider these methods. A general look at measurements taken on common baths gives a ready indication of what may be deduced from gravity and conductivity measurements.

Measurements of common baths were taken in terms of resistivity as ohm-cm at 25°C and gravity as °Bé at 25°C as shown in Table 19.

The same data of Table 19, in elaborated form, is presented in Table 20. A number of interesting deductions can readily be made:

The gravity is greatly influenced by the total concentration of salts, whereas the conductivity is not.

The four baths at the top of the table indicate that the conductivity is related approximately to the total concentration when all the baths are mildly alkaline.

The zinc and nickel sulfate baths are quite close to neutral and these baths are much less conductive even though the total concentrations are considerably higher. This confirms what is well known—

Table 19 Chemical Concentration, Resistivity, and Gravity of Various Plating Baths

	grams per liter	*ohm-cm at 25°C*	*°Bé at 25°C*
Brass Bath		12.4	10.9
copper cyanide	30		
zinc cyanide	10		
sodium cyanide	56		
sodium carbonate	30		
Cadmium Bath		7.3	11.3
cadmium oxide	30		
sodium cyanide	110		
Chromium Bath		2.0	21.6
chromium oxide	250		
sulfuric acid	2.5		
Acid Copper Bath		5.6	20.6
copper sulfate	200		
sulfuric acid	75		
Rochelle Copper Bath		14.3	11.0
copper cyanide	26		
sodium cyanide	35		
Rochelle salts	45		
sodium carbonate	30		
High Efficiency Sodium Cyanide Copper Bath		6.8	22.0
copper cyanide	120		
sodium cyanide	135		
sodium hydroxide	30		
High Efficiency Potassium Cyanide Copper Bath		5.9	13.9
copper cyanide	56		
potassium cyanide	85		
potassium hydroxide	41		

namely, that alkaline solutions are more conductive than neutral ones.

The remainder of the baths that are more strongly alkaline or acid are more conductive, but with a few exceptions the conductivity roughly indicates the concentration.

Chromic acid is an outstanding exception to the stronger baths. The conductivity of this bath by comparison is unusually high. This confirms what is also well known—that strong acids are highly conductive.

The lead fluoborate bath and the copper sulfate bath are quite conductive revealing the presence of strong acids.

A look at the resistivities (the reciprocal coductivities) gives a ready indication of what may be anticipated in terms of bath voltage. At

Table 19 *(continued)*

	grams per liter	*ohm-cm at 25°C*	*°Bé at 25°C*
Acid Lead Bath		4.5	22.6
lead fluoborate	105		
fluoboric acid	40		
Watts Nickel Bath		19.5	21.7
nickel sulfate	240		
nickel chloride	45		
boric acid	30		
sulfuric acid	0.23		
Cyanide Silver Bath		9.1	12.7
silver cyanide	40		
potassium cyanide	60		
potassium carbonate	60		
Alkaline Tin Bath		16.2	11.6
sodium stannate	90		
sodium hydroxide	8		
sodium acetate	15		
Cyanide Zinc Bath		7.7	12.0
zinc cyanide	60		
sodium cyanide	23		
sodium hydroxide	53		
Acid Zinc Bath		21.0	19.5
zinc sulfate	240		
ammonium chloride	15		
aluminum sulfate	30		
Alkaline Cleaner		19.0	5.8
sodium carbonate	30		
trisodium phosphate	15		
sodium hydroxide	8		

the same temperature (25°C) and current density, independent of polarization, the bath voltage is proportional to the resistivity. It is seen then that much higher voltage sources are expected for the neutral and moderately alkaline baths. However, the resistivity will greatly be decreased when the baths are heated.

The alkaline cleaner is interesting; the resistance of this bath is among the highest listed. We might conclude from this that a higher than usual voltage would be indicated to operate an electrolytic cleaner. From a comparison with the other baths a number of facts about the cleaner become apparent. The resistivity is high for two reasons. First, the total concentration is low; second, the cleaner is a medium duty cleaner with only 8 g/l of NaOH present.

Table 20

Bath	*Chemical strength*	*Conc., g/l*	*°Bé*	*sp. gr.*	*R, ohm-cm*	*C, mho/cm*
Sodium stannate	mod. alk.	113	11.6	1.087	16.2	0.062
Brass	mod. alk.	126	10.9	1.082	12.4	0.081
Copper cyanide	mod. alk.	136	11.0	1.083	14.3	0.070
Silver cyanide	mod. alk.	160	12.7	1.095	9.1	0.110
Zinc sulfate	sl. acid	285	19.5	1.115	21.0	0.048
Nickel sulfate	sl. acid	315	21.7	1.176	19.5	0.051
Alkaline cleaner	alk.	53	5.8	1.042	19.0	0.053
Zinc cyanide	alk.	136	12.0	1.090	7.7	0.130
Cadmium cyanide	alk.	140	11.3	1.085	7.3	0.137
Lead fluoborate	acid	145	22.6	1.185	4.5	0.222
Copper cyanide KHS	alk.	182	13.9	1.106	5.9	0.169
Chromic acid	st. acid	252	21.6	1.175	2.0	0.500
Copper sulfate	acid	275	20.6	1.166	5.6	0.179
Copper cyanide NaHS	alk.	285	22.0	1.179	6.8	0.147

The cleaner will be operated hot and can be compared to the hot alkaline tin bath. Actually, both of these baths are successfully operated on a 6-volt line.

The data of Table 20 indicate that gravity and conductivity readings (compared with the total bath concentration) and known data will reveal something about the bath characteristics. Conductivity is a potential control factor when baths are moderately alkaline or acidic. Slightly alkaline or acidic baths are better controlled by pH and strongly alkaline or acidic baths by titration. Gravity is a potentially useful factor in the majority of the baths.

Specific Gravity

A precise look at specific gravities of common salts shows how gravities differ as well as how some are quite similar.

Gravities of 10% solutions are listed in Table 21.

There is quite a difference in the gravities of the ammonium salts and the acids as compared to the others. Sodium and potassium are quite similar as are the metal salts at the bottom of the table. The table tells us that the gravity of a salt for which there is no data can be estimated by similarities if enough other data is available. The gravity of ferric sulfate on the other hand stands as a warning that anomalies exist.

Table 21 Gravities of 10% Solutions

Cation	*Anion*		
	Chloride	*Nitrate*	*Sulfate*
Ammonium	1.028	1.040	1.057
Hydrogen	1.047	1.054	1.066
Potassium	1.063	1.063	1.082
Sodium	1.071	1.067	1.092
Magnesium	1.082	1.076	1.103
Iron (ferric)	1.085	1.081	1.084
Zinc	1.082	1.086	1.107
Aluminum	1.090	1.081	1.106
Cadmium	1.091	1.087	1.102
Copper	1.096	1.088	1.107
Nickel	1.099	1.088	1.109

Table 22 Specific Gravity Constants

Chemicals	*Constant*
$ZnSO_4 \cdot 7H_2O$	0.000545
NH_4Cl	0.000309
$Al_2SO_4 \cdot 18H_2O$	0.000510
$CuSO_4 \cdot 5H_2O$	0.000622
H_2SO_4	0.000635
Na_2CO_3	0.000972
NaOH	0.001054
Na_2SO_4	0.000868
KOH	0.000876
K_2CO_3	0.000860
$NiSO_4 \cdot 6H_2O$	0.000577
$NiCl_2 \cdot 6H_2O$	0.000505
$Na_3PO_4 \cdot 12H_2O$	0.00037*
H_3BO_3	0.00060*
AgCN	0.00057*
CuCN	0.00057*
$Zn(CN)_2$	0.00057*
CdO	0.00057*
NaCN	0.00061*
KCN	0.00061*

* **First approximation.** The equation for a nickel bath is:
Sp. gr. $= 0.997 + NiSO_4 \cdot 6H_2O \times 0.000577 + NiCl_2 \cdot 6H_2O \times 0.000505 + H_3BO_3 \times 0.00060.$

Every plating bath has a maximum and a minimum allowable chemical concentration set by the control limits of the bath. The gravities are approximately additive and can be expressed by an additive equation when enough data is taken to estimate and confirm that specific gravity constants fit the equation. The equation takes the form:

$$\text{Sp. gr.} = 0.997 + A_1C_1 + A_2C_2 + A_3C_3 + \ldots$$

where A is the concentration of a chemical substance and C is a specific gravity constant.

The constants will change somewhat with a given substance from bath to bath and also to some extent with variation in total concentration. They have, however, been found to be quite reliable for control of many baths when sufficient gravity data has been taken.

A starting equation can be estimated from tested constants or tried first approximations as listed in Table 22.

By setting up similar equations the following calculated results were obtained as compared with the measured results of Table 20.

A little work with gravity data frequently reveals a simplification that fits well into the control scheme.

Table 23 Calculated and Measured Specific Gravities

	specific gravity at 25°C	
Bath	*meas.*	*calc.*
Brass	1.082	1.086
Cadmium	1.085	1.084
Copper sulfate	1.166	1.172
Rochelle copper	1.083	1.094
Copper cyanide NaHS	1.179	1.179
Copper cyanide KHS	1.106	1.120
Nickel	1.176	1.179
Silver cyanide	1.095	1.112
Zinc cyanide	1.090	1.104
Zinc sulfate	1.155	1.151
Alkaline cleaner	1.042	1.043

Conductivity

The conductivities and resistivities of various chemicals are listed in Table 24.

The values of Table 24 demonstrate that the acids and the alkalis are highly conductive as compared to the salts, confirming the observations of Table 20.

Current is carried across a solution by the individual ions so the conductivity is dependent on the percentage of ionization and the conducting capacity of the ion. As a basis for comparison this can be

Table 24 Conductivites and Resistivities of Various Equivalent Chemical Solutions at 20°C

Chemical	*g/l*	*R, ohm-cm*	*C, mho/cm*
HNO_3	63	3.2	0.31
HCl	36	3.3	0.30
H_2SO_4	49	5.0	0.20
KOH	56	5.4	0.18
NaOH	40	6.4	0.16
KCl	74	10.2	0.098
NH_4Cl	53	10.3	0.097
NaCl	58	13.4	0.075
K_2CO_3	69	14.1	0.071
$AgNO_3$	170	14.8	0.067
Na_2CO_3	53	22.0	0.045
$CuSO_4 \cdot 5H_2O$	125	38.7	0.026

expressed as a theoretical value in mhos per chemical equivalent conducted by each type of ion at a distance of one centimeter and at infinite dilution. These values are useful to calculate the conductivities of dilute solutions but our interest here is only in the relative values to demonstrate the relative conducting capacity of various ions (Table 25).

Since at infinite dilution ionization is complete, the conductivities are additive. Thus, the conductivity of KCl at 25°C would be

$$\mathrm{K} + \mathrm{Cl} = 74 + 75 = 149$$

and

$$\mathrm{H} + \mathrm{Cl} = 350 + 75 = 425$$

Table 25 Equivalent Conductance of the Ions at 25°C and 100°C

	Conductivities	
Ion	*25°C*	*100°C*
K	74	206
Na	51	155
NH_4	74	207
Ag	63	188
Cl	75	207
$\frac{1}{2}SO_4$	79	234
H	350	644
OH	192	439

showing how H contributes to the conductivity.

There are two points of interest in Table 25. The conductivity of the hydrogen ion and the hydroxyl ion are much greater than the other ions. Second, the conductivity increases substantially with increase in temperature.

Voltage

When voltage is given as a part of a bath formulation it indicates the voltage that should be available to put a bath in operation. This may or may not include an excess that is expended in the rheostat as an essential to electrical control.

The bath voltage depends on the potential at the anode, the potential at the cathode, sizes and spacing of the anode and cathode, the current applied, and the resistivity of the bath. The potentials at the electrodes are small except where polarization is essential, as in the alkaline tin bath. The other factors mentioned are related to resistivity.

Most plating baths operate well from a 6 to 9 volt line and there is no particular voltage problem. However, if plating is to be done at high current densities or at large anode-cathode distances it is convenient to be able to estimate the bath voltage. This can be done when the resistivity at the operating temperature is known or can be estimated.

An approximation of the bath voltage can be made from the equation

$$E = \frac{IRL}{A}$$

where E is the voltage, I the current, L the anode-cathode distance, A the cross sectional area, and R the resistivity.

Some assumptions must be made. For the purpose of an estimate it is assumed that the volume through which the current flows is defined by straight lines from the anode to the cathode. This is far from true but it represents the maximum voltage condition.

For example: The resistivity of a Watts nickel bath at 40°C is 15 ohm-cm. Assume 1 sq ft of anode area, the same cathode area, and a 6-inch anode-to-cathode distance. At a current density of 30 the values are: $I=30$, $R=15$, $L=15.3$ cm, $A=930$ sq cm. Therefore,

$$E = \frac{IRL}{A} = \frac{30 \times 15 \times 15.3}{930} = 7.4 \text{ v}$$

Electrode polarization of about 1 volt is expected; therefore, a voltage source of at least 9 volts would be recommended. Actually, the tank voltage would be somewhat less but a voltage of about 2 should be allowed to operate the rheostat and a line loss of 1 volt might be expected. With a few simple measurements or other data these voltages also can be estimated.

Bath voltages calculated from the resistivites of Table 19 give an indication of what might be expected of the various baths. Table 26 shows the probable electrode polarizations of the baths and calculated voltages for current densities of 10 and 100 at an anode-cathode distance of 1 foot and operation at 77°F.

By reference to the table an estimate can be made for a wide range of plating conditions. In general, the conductivity of a bath will increase about 2% per °C. This generalization may be applied to moderate temperature changes but for a large difference the resistivity at or near the operating temperature ought to be measured.

Table 26 shows immediately that quite high voltages result at extreme conditions. A voltage of 65 is required for a Watts bath at room temperature, 100 asf and electrode distance of 1 foot. By increasing the temperature, by using a chloride bath, and by reducing the

Table 26 Approximate Polarization and Calculated Voltages for Common Plating Baths

Bath	Estimated electrode polarization	*Voltage at current density* 10 asf	100 asf
Brass	0.3–0.7	4.5	41
Cadmium	0.5–1.0	3.0	24
Chromium	1.0–2.0	2.5	7
Copper sulfate	0.2–0.5	2.2	19
Rochelle copper	1.0	5.7	48
Copper cyanide NaHS	1.0	3.2	23
Copper cyanide KHS	1.0	2.8	20
Lead fluoborate	0.2–0.5	1.8	15
Watts nickel	1.0–1.5	7.5	65
Silver cyanide	0.5–1.0	3.5	31
Alkaline tin	1.0–3.0	7.0	53
Zinc cyanide	1.0	3.5	26
Zinc sulfate	0.1–0.2	7.0	69
Alkaline cleaner	0.5	6.7	63

electrode distance to ½ inch, a nickel bath may be operated at 500 asf from a 6 volt line. Also given is 63 volts for an alkaline cleaner at 100 asf; yet, many cleaners are operated at this current density from a 6 volt line. This is accomplished by using a high temperature and a high caustic content.

34. ELECTROPLATED ALLOYS

In normal plating baths, only one metal is deposited and the electrodeposits are usually of high purity. However, if noble-metal impurities enter the plating bath, they will be codeposited with the major metal. Such impurities are a common source of trouble and result in rough, porous, and off-color deposits. The behavior of these noble metals in a plating bath leads one to believe that it is not feasible to deposit more than one metal at the same time. In contradiction to this belief, there are cases where impurities are codeposited with the major metal without apparent interference. Small amounts of iron, for example, will codeposit with nickel in a Watts nickel bath without detection. Larger amounts will cause the deposit to be dull or burned. Appreciable amounts of cobalt may be codeposited with nickel. In fact, bright deposits are readily obtained by codeposition of nickel-cobalt alloys in the presence of addition agents.[1,2] Iron-cobalt and nickel may all be deposited simultaneously in a variety of proportions.

Iron, nickel, and cobalt serve well to illustrate the requirements for alloy plating. In order to codeposit metals, they must be compatible —that is, they must be compatible both from the metallurgical and the electrochemical standpoint.

Metallurgically, they must be metals that will form solid solutions or intermetallic compounds. The minor metal must fit into the lattice of the parent metal. This means that the most successfully plated alloys are the same ones that are successfully prepared by casting procedures. Examples of alloys successfully prepared by both procedures are brass, bronze, and solder. It must be kept in mind, however, that electroplating is a cold process as compared to casting. Therefore, the equilibrium at the plating temperature will be far from that at which alloys are formed from a melt. In order to predict the possible results to be obtained by electrodepositing an alloy, the phase diagram should be examined in the low-temperature region. However, predictions cannot be made with certainty because electrodeposits are often finely crystalline and fine crystal size may promote greater solubility than estimated from a phase diagram.

The physical chemist knows that very small crystals are more soluble than large ones. When the surface area is large compared with the mass, deviations from the phase diagram may be expected. Lead, for example, has been codeposited with silver as a solid solution containing 10% lead, which is much higher than expected according to the phase diagram. Codeposits form solid solutions and intermetallic compounds, but the structures will not be similar to cast structures. First, they will be more finely crystalline and second, certain structures will not be obtained at all. A eutectic structure cannot be obtained because it is formed from a melt and no parallel to such a freezing process takes place during deposition. Yet, in some cases, the structure of electrodeposits may approach the cast structures. If an alloy is cast, then cold-worked and annealed in order to obtain a fine structure and to attain low-temperature equilibrium, it may approach the structure obtained from an electrodeposit annealed at the same low temperature.

The points of comparison between cast and electrodeposited material are as follows:

(1) The most successfully prepared alloys are those formed from metals that have an affinity for each other, but different results may be expected with different processes of formation.

(2) The alloying metals must be compatible from the electrochemical standpoint.

(3) Iron, cobalt, and nickel are similar electrochemically. They all tend to deposit at approximately the same voltage. Therefore, codeposits of these metals are readily obtained. Since sound alloys can be prepared from these metals, an alloy deposit is obtained. An alloy deposit is, in this sense, defined as a sound deposit. Codeposits may be formed merely because ions of two metals are discharged at the cathode simultaneously, but the structure will be weak if the metals do not have an affinity for each other.

Iron, cobalt, and nickel codeposit because they have a similar deposition potential. By referring to the electrochemical series, it will be seen that the equilibrium potentials of the metals cover a wide range. If the equilibrium potentials are far apart, the deposition potentials in solutions containing simple ions probably will not be favorable. Therefore, the number of alloys successfully deposited from acid solutions is considerably less than can be expected to be deposited from the complex-ion solutions. The potentials of lead and tin are close together so that they are expected to codeposit. Actually lead-tin alloys can be deposited from acid solutions. However, tin does not codeposit readily unless addition agents are used.

Alloy plating from solutions of simple ions is limited because of the deposition potentials of the metals. These potentials can be changed somewhat by changing concentration, acidity, temperature, current density, etc. But there is nothing that can be done to codeposit an active metal like zinc with copper from acid solutions. Small amounts of a nobler metal may be deposited, such as gold with copper and copper with lead, but here again the codeposition is limited to small amounts. The reason for this is that the noble-metal plating rate depends on the rate at which it reaches the cathode. For instance, copper will codeposit with lead as fast as the copper diffuses to the cathode. If a lead cathode is used, the copper will continue to deposit even though the current is not flowing. In fact, copper will deposit on a lead anode while lead is going into solution under the influence of current. Thus the copper is not under the influence of the usual plating variables in the same manner as lead. If the acid bath contains large amounts of copper, it will not be possible to codeposit lead. We prefer to call a plating system of this type a "noble-metal system."

In any alloy system, one of the metals acts as a noble metal—in other words, one metal tends to deposit more readily than the other. It is necessary then to use the term noble metal even in the potential system.

The noble-metal system is recognized by the fact that the plating rate of the noble metal is primarily dependent on the concentration of this metal in the bath. In other words, it is independent of the concentration ratio of the two metals.

The potential system is recognized by the fact that the percentage of metal obtained in the deposit is dependent on the ratio of the metals in the bath.

The noble metal system is limited, in solutions of simple ions, to small amounts of the noble metal in the deposit.

Most acid solutions provide simple ions and the deposition potential depends on the ionic concentration. If the ionic concentration can be changed, then the tendency to plate (the potential) can also be changed, i.e. can be controlled. The potential can be changed about 0.06 volt by every 10 times change in concentration, so that if the potentials of two metals are wide apart it is not feasible to bring them together by change in concentration alone.

Ionic concentrations can be greatly changed by the formation of complex ions. These complex ions are most conveniently formed in cyanide solutions for plating purposes. For instance, the following complex is formed in a silver cyanide bath:

$$KCN + AgCN = KAg(CN)_2$$

This complex ionizes as follows:

$$KAg(CN)_2 = K^+ + Ag(CN)^-{}_2$$

In this primary ionization no silver ions are formed. A secondary ionization takes place that gives rise to a small amount of silver ions:

$$Ag(CN)^-{}_2 = Ag^+ + 2CN^-$$

Cadmium, zinc, gold, copper, iron, and a number of other metals form cyanide complexes. Tin and lead form complexes with caustic soda. If caustic soda is used to dissolve a tin salt, the tin-ion concentration will be so greatly reduced that tin will not deposit from a cold solution. Iron forms a cyanide complex that is so stable ionically that iron will not deposit on passing current across the bath.

By the formation of complexes with cyanide and caustic soda, the potentials of the metals can be greatly changed. If a copper salt and sodium cyanide are added to an akaline tin bath, copper can be deposited. If the solution is operated cold, copper and hydrogen will appear at the cathode. The deposition potentials of copper and hydrogen then must be approximately the same under these conditions. If the solution is heated and a considerable excess of cyanide is added, then tin and hydrogen will appear at the cathode. Copper will not deposit even though a high concentration is present (with a great excess of free cyanide). By proper adjustment of conditions, tin and copper can be codeposited without depositing hydrogen. In fact, by proper adjustment of one, any two or all three ions can be deposited at the same time.

By the formation of complexes, it is possible to codeposit alloys that cannot be deposited from acid solutions. Tin and copper can be codeposited in all proportions from complex-ion solutions. Tin-copper-zinc alloys can also be deposited from these solutions.

Cyanides are the most satisfactory for formation of complexes and for successful alloy plating since many of the present plating baths are cyanide baths. By addition of caustic soda to the bath, further control of the deposition potentials is possible. Zinc and indium form complexes with both cyanide and caustic soda.

By the use of cyanide baths, copper can be codeposited with tin and zinc and silver can be codeposited with lead and cadmium. Deposition of many alloys that are as yet unexplored is possible although hundreds of alloys have been codeposited.

Although many alloys have been deposited, only a few are used commercially. Brass is deposited quite widely. Solder, speculum metal, and copper-tin-zinc alloy are deposited to a lesser extent.

There is interest in alloy plating, but it is not widely used in comparison with plating of the pure metals. The primary reason for this is that the development of alloy baths is difficult. There is more to the problem than merely bringing the potentials of the metals together. In order to be commercially successful, the bath must be stable, the salts used must have an appreciable solubility, the deposit obtained must be sound, and continuous control of the bath must be possible. The alloy must deposit at an appreciable rate, at the usual current densities, and at a sufficiently high cathode efficiency. Bath balance must be possible to maintain control. This problem includes the development of soluble anodes.

Wider use of alloy deposits is dependent on the demand. They are widely used and are successful as coatings for other metals. It remains to more fully demonstrate the value of alloy deposits. Among the advantages of alloy deposition, the following may be mentioned:

1. *Decorative Value*. Bright alloys can be deposited such as nickel-cobalt and copper-tin-zinc.

2. *Economy*. Certain alloys will be cheaper than pure metals used for the same purpose.

3. *Corrosion*. It is well known that the corrosion resistance of many alloys is greater than that of pure metals.

4. *Engineering*. Lead alloys are being successfully used as the running surface for bearings.

5. *Variable Properties*. Where an application is dependent on the properties of the coating, some control of the properties is possible by the application of alloys.

It is important that the metallurgist and the engineer consider alloy deposits where coatings are used. By realization of how the properties of alloys may be applied to coatings, the use of alloy plating will spread. No one knows better than the metallurgist how the mechanical properties of a coating may be changed by variation of the alloy composition. Wear resistance, hardness and ductility may be changed depending on the demand for greater wear resistance or better formability.

The brass bath is a good example of how a problem was solved by the use of an alloy deposit. In the rubber-tire industry, it was desirable to bond rubber to steel wires. After considerable experimentation, it was found that rubber could be bonded to brass and that thin coatings of brass could be bonded to steel by electroplating. Thus, the rubber could be bonded to steel by the use of brass as a bonding medium. This application greatly stimulated the development of the brass-plating bath.

Electrochemistry of an Alloy Bath

An alloy-plating bath differs from a single-metal bath in that greater control is usually required. More than one metal is present in the bath and each metal will respond differently or the alloy will respond to a greater extent to changes in bath conditions. Since limits are placed on the percentage of metal in the deposit, the bath variables must be controlled to operate within these limits.

If we have two metals depositing at the same time, one will tend to deposit more readily than the other. This metal is called the noble metal. If there is an appreciable difference in the tendency of the two metals to deposit, the noble metal will deposit exclusively at low current density. As the current density is increased and as the diffusion layer at the cathode film thickens, it becomes more difficult for the noble metal to deposit in competition with the second metal present. Any variable that tends to increase polarization or to thicken the cathode film will decrease the relative ability of the noble metal to deposit. Polarization is decreased, the cathode film is made thinner, and the noble metal deposits more readily as a result of the following changes:

1. Decrease in current density.
2. Increase in agitation.
3. Increase in temperature.

The variables are the same as those that fix the limiting current density in a single-metal bath.

If the limiting current density is low, it may be raised by raising the temperature or increasing the agitation. The extent of the change in

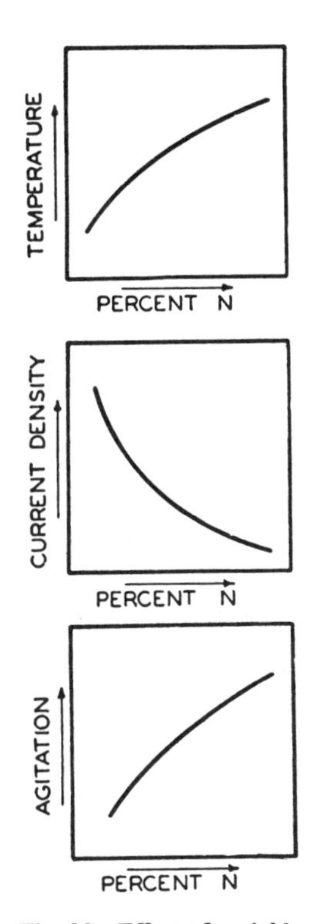

Fig. 23 Effect of variables on alloy deposits (N=noble metal in deposit).

noble-metal content of the deposit with the change in bath variables depends on the difference in tendency of the two metals to deposit. If this difference in tendency is great, then control may be difficult. If the difference is small the nobility may even reverse with change in conditions so that the second metal becomes more noble.

These rules are general and may not hold where temperature or chemical concentration affects the ionization of one metal more than that of another. By such variables, nobility may be reversed and the point of reversal is the ideal point or the point at which both metals will respond to the same extent to a change in current density. (see Fig. 23.)

There has been a great interest in alloy plating from the early days of electroplating and there have been many general studies made to advance alloy plating and extend the properties of electroplates by this means. It is now apparent that there is no simple set of principles that can be applied to the practice of alloy plating.[3] However, each alloy bath responds in some known ways to temperature, concentration, agitation, and current just as do the single metals. The response of competing ions to these variables has been established by experimentation in the practical baths. Thus, from the economic standpoint each commercial alloy bath becomes just another plating bath that requires a little more attention, but produces deposits with properties unattainable from single metals.

REFERENCES

1. C. B. F. Young and Clifford Struck, *Trans. Electrochem. Soc.*, **89,** 383 (1946).
2. L. Weisberg, *Trans. Electrochem. Soc.*, **73,** 435 (1938).
3. A. Brenner *Electrodeposition of Alloys: Principles and Practice*, 2 volumes, Academic Press (1963).

35. LAYER PLATING

Two ideals of an electrodeposit are corrosion resistance and freedom from pores. Many coatings satisfy the first ideal but the second generally is not attained when deposits are less than 0.002 inch thick. The problem of porosity is, however, alleviated by plating several layers of metal. Something is gained merely by interrupting the plating current or by periodic reverse plating, so that plating will bridge the pores and interupt the porous path by which moisture can penetrate to the substrate.

The classical example of layer plating is chromium/nickel/copper. The copper is deposited, then buffed, to flow the pores closed and to provide a bright surface for subsequent nickel. The nickel is then applied to protect the basis metal from corrosion. Chromium is deposited on the nickel to protect the surface rather than the substrate and to preserve a bright stain-resistant finish.

Substantial improvements have been made in chromium/nickel corrosion resistance by plating layers of duplex and triplex nickel. These processes depend on plating layers of nickel from different baths such that one layer of nickel is sacrificial to another, thus protecting the underlying steel.

A number of commercial methods have been developed that depend on plating of several layers and then heating to form diffusion alloys. Layer plating of this type offers some potential advantages over alloy plating in that a range of alloy properties can be produced that extend from the surface to the substrate. A good example is the process known as *Corronizing*. It consists of an undercoat of nickel plated with zinc or tin and then heated to form an alloy coating.[1] When the zinc-nickel layers are heated to 700°F a protective coating is produced that consists of layers of alloys. The alloy at the surface is 80% zinc while next to the steel it is pure nickel. The coatings have been used to protect wire screen, oil cans, and outboard motors.

Military Specification MIL-P-23403[2] on tin-cadmium plating recognizes that a diffusion alloy and an electroplated alloy may be regarded as equivalent, at least in this instance, by the following statement: "The plating may be deposited as an alloy or tin and cadmium may be deposited separately and diffused." Tin-cadmium deposits, to satisfy this specification, must contain from 25 to 50% tin.

Lead-indium alloys were deposited, to produce running surfaces for bearings, by depositing 4% indium over 0.001 inch of lead and then diffusing one hour at 320°F. An alloy that served the same purpose was formed by diffusing 10% tin into 0.001 inch lead at 300°F.

There is a wealth of background and many possibilities in the field of layer plating and in the field of diffusion alloys.

Diffusion Coatings

Many metallurgical methods of producing diffusion coatings have been developed.[3] These general nonelectrolytic methods are of interest because of what they suggest might be applied to layer plating or to diffusion of a surface applied metal. Steel alone is treated with aluminum, carbon, chromium, nitrogen, silicon, tin, and zinc by calorizing, carburizing, chromizing, nitriding, ihrigizing (siliconizing), stannizing, and sherardizing. The various methods may be carried out with powdered metal, molten metal, a salt, or a gas, each method producing a surface with properties different from those of the basis metal. The object of these treatments is usually to produce a more wear resistant or a more corrosion resistant surface. These thin coatings then compete directly with electrodeposits. The properties of the coatings depend on the type and the extent of the alloy formation.

Table 27 Possible Alloy Formation from Binary Phase Diagrams

	Pb	*Ag*	*Cd*	*Cu*	*Sn*	*Al*
Fe	1	1	1	2	3	3
Ni	1	1	2	2	2–3	3
Zn	1	2–3	2	2	1	2
Pb	2	2	2	2	2	1
Sn	2	2–3	2	2–3	2	2
Ag	2	2	2–3	2	2–3	2–3
Cu	2	2	3	2	2–3	2–3

1 = No appreciable alloy formation.
2 = Solid solution.
3 = Compound formation.

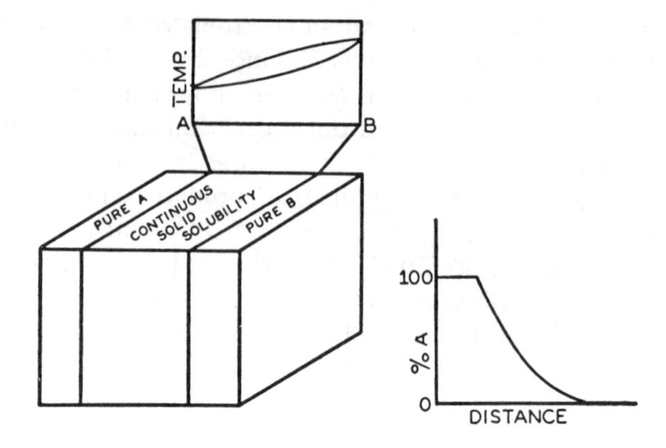

Fig. 24 Continuous solid-solubility.

As we have seen, alloys can be formed by plating a metal on another and then subjecting to heat. The number of possible alloys that can be formed is great. Table 27 shows expected alloy formation from a study of a number of binary phase diagrams.

Solid-solubility or compound formation must be possible for diffusion to take place. When diffusion does occur, the phases formed will be as indicated in the low-temperature region of the phase diagram and they will appear in the same order. All the single phases in a binary diagram will appear as layers but the two phase regions will only appear as lines separating the single phases.

The following diagrammatic cases will illustrate the relationship between a phase diagram and a diffusion alloy.

Solid-Solubility

When two metals form a continuous series of solid solutions, one of the metals will diffuse into the other. Thus, an alloy zone will be formed that will range from 0 to 100% metal (Fig. 24). If diffusion continues until the pure metals are used up, then the alloy zone will homogenize and form an alloy of a percentage depending upon the amount of each metal originally present.

When two metals form solid solutions of limited solid-solubility, the method of diffusion will be similar to that for continuous solid-solubility except that two phases will be formed (Fig. 25).

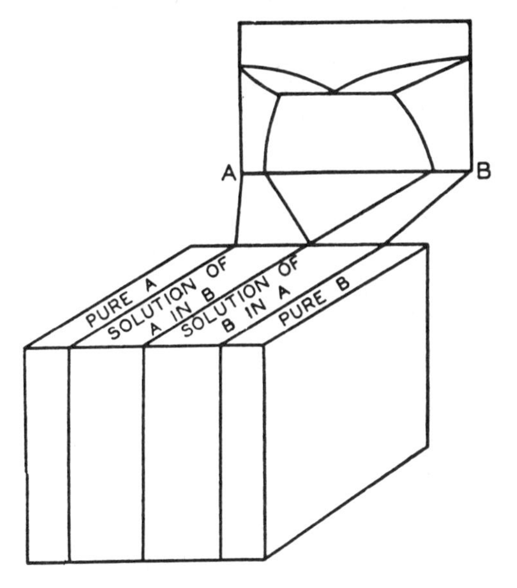

Fig. 25 Limited solid solubility.

Compound Formation

If compounds or intermediate phases are indicated on the phase diagram, then these will also be formed on diffusion (as in Fig. 26).

The phases that are formed will appear in the order indicated on the phase diagram, but the quantity formed can only be determined by experiment or through a knowledge of diffusion rates. However, after the phases are formed, the ratio of thickness of the diffusion layers will remain the same as long as some of each of the original metals is present. After one of the metals is completely used up, homogenization will tend to take place. If homogenization does take place, then only one phase will remain, but this condition may take a very long time to occur. Small amounts of the phases are formed very readily, but as the thickness of the diffusion layers grows, any further increase in growth is slow. The thickness of the diffusion layer will be increased approximately as the square root of the time; thus, to double the thickness, the diffusion time will have to be increased four times.

Diffusion can take place at temperatures appreciably below the solidus on the phase diagram. Low-melting metals, like tin and indium, may even diffuse slowly into some metals, such as lead, at room

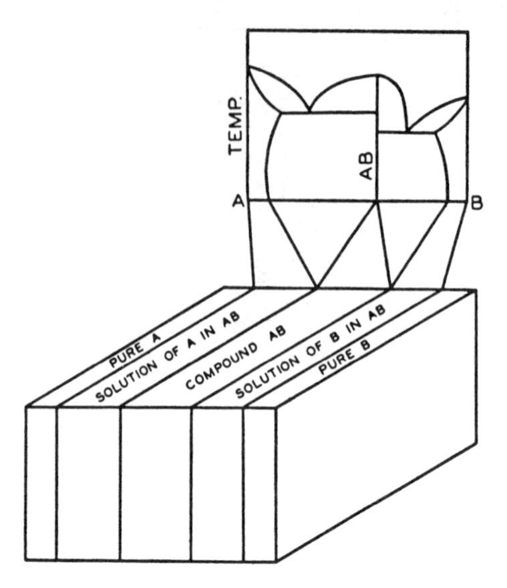

Fig. 26 Compound formation.

temperature. Of course, the rate of diffusion under these conditions is slow, but it can be greatly accelerated by increasing the temperature. The diffusion rate will increase as the diffusion temperature approaches the melting point of the lower-melting of the two metals. If the solid-solubility of the two metals concerned is broad, then diffusion is likely to go on at a fair rate at elevated temperatures and a broad diffusion zone is possible. If three metals are diffused at the same time, then diffusion to form a ternary system may be more rapid than with two metals. This would be particularly true if the three metals were appreciably soluble in one another and would melt at a lower temperature than any of the three possible binary combinations.

Many new alloy combinations are possible by taking advantage of diffusion alloys produced from electrodeposits. The alloys that result may be homogenized or they may be in layers. If they are in layers, the properties will be different from those of alloys produced by casting or by electroplating.

Further study is required on diffusion in general and on diffusion of electrodeposited layers, but it is certain that many alloys, such as copper-zinc, copper-tin, copper-nickel, silver-zinc, silver-tin, lead-base, tin-base, zinc-base and many others, can be produced.

REFERENCES

1. R. Rimbach, *Metal Finishing*, **39,** 360 (1941).
2. MIL-P-23403, *Plating: Tin-Cadmium* (*Electrodeposited*).
3. F. N. Rhines, "Surface Treatment of Metals," American Society for Metals, p. 122 (1941).

36. APPLICATIONS OF ELECTROPLATING

The early applications of electroplating were primarily ornamental, and even today the majority of electroplates are applied essentially for decoration.

More than 100 years ago, silver was electrodeposited on metals as a cheap method for producing silver-clad objects. These objects were valued because people have always surrounded themselves with objects that are pleasant to look at. The silver-plated objects were either partially useful and partially ornamental, such as tableware, or they were wholly ornamental, such as jewelry.

Today, we find electroplates applied for ornamental purposes on tableware, hardware, electrical fixtures, automobile trim, and a host of other objects. In fact, it is almost impossible to go through a day without using a score of electroplated articles.

The plates on inexpensive objects such as paper clips and safety pins must be produced at very low cost. In cases such as these, electroplating must produce deposits of satisfactory appearance, but economy has entered into the production.

Economic Applications

It is difficult to separate economy from ornamentation since it is often necessary to impart a pleasing appearance to an object at low cost. Even for the more expensive objects, such as tableware, electroplates can be produced far more cheaply than sterling silver and yet the plate can be produced to last a lifetime.

Some applications of electroplating, however, are for purely economic purposes. Plating that is applied almost entirely for corrosion resistance is in this class. For example, zinc, lead, or tin may be applied to sheet metal to insure sufficient life under conditions of exposure to moisture, weathering, or corrosive liquids.

Engineering Applications

Engineering applications are those in which a better job is done by plating than by some other means. Plating in this class is less widely used, but there are a number of important applications. These include salvage of improperly machined parts, build-up of worn parts, production of wear-resistant surfaces, bearings, electrical contacts, electroformed parts, and selective plating prior to carburizing.

The list of applications shown in Table 28 indicates the potentialities of electroplating.

When electroplating applications are of a miscellaneous nature, various problems are presented. For example, if it is required to plate a part for temporary protection against corrosion, considerable experimentation may be in order. First of all, decisions must be made regarding the probable exposure life of the part. The exposure conditions may be varied; for instance, the parts may be replacement parts for a mechanical device and they may lie on the distributor's or salesman's shelf for several years. The customer usually is familiar with the conditions to which the parts will be exposed and can satisfactorily set up the requirements for the product.

Another factor to be considered is the importance of the appearance of the electroplated part. Attractively finished parts have greater sales appeal even if the parts go in the interior of an appliance where they are very seldom seen.

In addition to corrosion resistance, appearance, and economy, the plating process has to be compatible with the rest of the operations going into the manufacture of the part. If, for example, part of the metal is machined away during manufacture and a part of the deposit is returned for remelting with the metal scrap, the deposited metal must not be objectionable in such a manner as to contaminate the alloy. If the deposited metal is detrimental to scrap recovery it will be necessary to develop a means of removal of the metal prior to reclaiming the scrap material.

The plating process itself must also be considered. Is the process adaptable to the part? What will the cleaning and pickling steps do to the part? Do recessed areas have to be covered? Is selective plating called for? Can the deposit be bonded to the metal or alloy in question? These qusetions and many others can only be answered by complete understanding of the problem under consideration.

Table 28 Typical Applications of Electroplating

Use	*Electroplate*
Automobile trim	Chromium (with undercoat of nickel and copper)
Battery parts	Lead
	Lead-tin
Bearings	Silver and silver alloys
	Lead alloys
	Tin alloys
Break-in (pistons)	Tin
Build-up of worn parts	Chromium
	Nickel
	Iron
Containers(food)	Tin
(chemicals and oil)	Lead and lead-tin
Corrosion resistance (weather)	Zinc
	Lead
	Cadmium
Corrosion resistance (chemical)	Specific for the chemical
Corrosion resistance (oil)	Lead and lead-tin
	Cadmium
	Copper-base alloys
	Nickel
Eating utensils	Silver
	Nickel
Electrical contacts	Silver
	Silver-rhodium
	Gold
Electroforming (general)	Copper (acid)
	Nickel
	Iron
Electroforming (sheet)	Copper
	Nickel
Electroforming (tubes)	Nickel
Electroforming (screen)	Nickel
	Copper
Electronic equipment	Silver
	Silver and rhodium
Electrotypes	Copper (acid)
Gun barrels	Chromium
Laboratory equipment	Silver
	Gold
	Nickel (copper undercoat)
Machines (food processing)	Tin
	Nickel
Machines (laundries)	Cadmium
Moisture resistance (general)	Zinc
(marine)	Cadmium
(steel in concrete)	Cadmium
(structural)	Lead
(washing equipment)	Cadmium
Molds (lining)	Chromium
Musical instruments	Silver
	Gold

Table 28 *(continued)*

Use	*Electroplate*
Nuts and bolts	Lead
	Zinc
	Cadmium
	Nickel
	Chromium
Ornamental	Silver
	Gold
	Rhodium
	Nickel
	Chromium
Pins, paper clips, etc.	Tin
	Copper
Plumbing fixtures	Chromium (nickel and copper undercoat)
Printing plates	Nickel
	Iron
Reflectors (general)	Cobalt
	Lacquer over silver
	Chromium
	Speculum
Reflectors (infrared)	Gold
Refrigeration coils	Tin
Rubber adhesion	Brass
Salvage of worn parts (specific)	Depending on application
(general)	Chromium
	Nickel
	Iron
Selective carburizing	Copper
	Bronze
Sheet metal (food containers)	Tin
(roofing)	Lead
	Lead-tin alloy
(outdoor exposure)	Zinc
Strike plating baths (general)	Cyanide copper
(silver-plating)	Cyanide silver
(cadmium adhesion)	Acid cadmium
Tarnish resistance	Chromium
	Tin
	Gold
	Platinum
	Chromium (nickel and copper)
	Nickel (copper)
Wear (hardness and low coefficient of friction)	Chromium
(hardness and electrical conductivity)	Rhodium over silver
(limited wear)	Nickel
	Iron
(cylinder liners)	Porous chromium
Wire (steel)	Zinc
(copper)	Tin

Plating can be used to a greater advantage through a knowledge of the properties of the individual metals and a knowledge of the characteristics of the plating baths. For instance, chromium is widely used to build up worn machine parts. Salvage of tools and dies by such a process makes it possible to reclaim old tools for a fraction of the cost of new ones. But heavy chromium deposits are not cheap to produce. Hard nickel and iron deposits will often serve for tool salvage just as well as chromium and their application should be given every consideration. These baths are easier to operate, the racking problems are not as difficult, and rapid plating rates are more easily obtained.

Plating of thin deposits may be applied for many purposes in a large plant. Machined parts often rust if they wait more than a few days for the next operation, or they may even rust in a matter of hours if the temperature and humidity are high. The application of a thin electrodeposit may protect the parts from rust and the savings realized in reduction of scrap will more than pay for the extra step involved.

Dip-coatings of certain organic substances are more economical than electroplated coatings for temporary protection, but it is possible that plating can be applied at such a phase of the process as to be of more permanent value. Also, the plated coating may make subsequent machining or handling operations easier. A coating may be selected that will not retain chips as readily as an oiled surface, or one may be selected that is easily soldered in a subsequent manufacturing step. When necessary, the deposit can be removed by a stripping method later in the process of manufacture. Plating may even be used as a means of identification of special parts. In large-scale manufacturing, special parts are occasionally run and there is a great danger of getting them mixed with regular production. For example, if all regular production were plated with zinc and special production with copper, the problem of segregation of the two parts would be simplified.

When a new or different plating bath is selected it is good practice to prepare several plating baths and run plating-range tests. Even if the baths are well known, nothing reveals the characteristics better than simple plating tests at various temperatures over a range of current densities. If the application is of sufficient importance, the tests may be extended to determine the effect of variation in chemical concentration and pH of the baths. This procedure of thoroughly studying the plating baths has further advantage in that the original plating-range standards can be used for future bath control. By using the plating-range test, small baths can be prepared and maintained. If they have sufficient life, it is easier to discard them and prepare new baths than to control them by analytical methods. However, some baths cannot be used fresh. Nickel, for instance, usually requires some working or purification before use. The same steps will be required for small laboratory baths as for production baths, except that the small baths can be made, filtered, and electrolyzed in a matter of hours, whereas it may take days to do the same on a large scale.

The next problem in experimental plating is to plate some actual production pieces as this brings in the cleaning, etching, and testing steps. In every new plating problem, a test for adhesion is probably the first step. For very thin deposits up to 0.05 mil, used for temporary protection, poor adhesion may be permissible, but in practically all other cases, it is necessary to obtain bond.

A good way to test for bond in the laboratory is to deposit 10 to 30 mils of metal and then attempt to remove the deposit by mechanical means. Bending, twisting, and pounding tests are often used, and a hammer and chisel can be used to learn much about heavy deposits.

The last stage of investigation consists of some type of test to predict the probable behavior in service. Such tests include corrosion, wear, and hardness, and are used as a basis for recommendation. If the recommendation leads to plant trial, then a pilot plant can be set up to determine adaptability of the plating process and service tests will determine the acceptability of the product.

If an unusual plating problem emerges, the possibility of baths other than the common plating baths should be considered. Under the proper conditions, other metals can be deposited. If nickel and iron are candidates in the problem, cobalt should certainly be considered. If a very hard surface is required, a tungsten alloy may be worthy of consideration. Arsenic, antimony, and aluminum can be deposited for experimental work. Many alloys and diffusion coatings are possible. It should be kept in mind, however, that these processes are not usable on a large scale without considerable experimental work and also that an experimental bath may fail in the laboratory (when applied to service) for what appears to be a minor reason. An experimental bath may fail because the bath is not stable, the plating range is narrow, the anodes are not soluble, or even that analytical methods are not available or cannot be developed. If an addition agent problem exists, hundreds of tests may be required.

Experimentation with new and unusual plating baths should only be

Table 29 Properties of Common Plating Baths

Metal	*Bath*	*Plating Range*	*Plating Rate*	*Throwing Power*	*Hardness*	*Corrosion Resistance*	*Control*
Brass	Cyanide	Wide	Slow	Good	Fair	Good	Careful
Bronze	Cyanide	Fair	Fair	Good	Fair to high	Good	Careful
Cadmium	Cyanide	Wide	Slow	Good	Low	Poor	Average
Chromium	Acid	Narrow	Fair	Poor	High	Excellent	Easy
Copper	Acid	Narrow	High	Poor	Fair	Tarnishes	Easy
Copper	Cyanide	Wide	Low	Good	Fair	Tarnishes	Average
Copper	High efficiency cyanide	Wide	Fair	Good	Fair	Tarnishes	Average
Lead	Acid	Narrow	High	Poor	Low	Oxidizes	Careful
Nickel	Acid	Fair	High	Poor	Fair to high	Good	Average
Silver	Cyanide	Wide	High	Good	Fair	Fair	Easy
Tin	Acid	Narrow	High	Poor	Low	Good	Difficult
Tin	Alkaline	Wide	Low	Excellent	Low	Good	Easy
Zinc	Acid	Narrow	High	Poor	Low	Poor	Average
Zinc	Cyanide	Wide	Low	Good	Low	Poor	Average

undertaken if it is permissible to assume an extensive laboratory and pilot-plant investigation.

Most plating problems should be attacked by starting with the standard common plating baths. Some of the properties and characteristics of these baths are listed in table 29 as a guide for selection.

An interesting application that points up a number of unique problems is the use of an electroplate as a stop-off, an example being copper plating for selective carburization.

Copper Plating for Selective Carburization

Certain applications call for areas of machine parts to be hard while other areas must remain relatively soft. The hard areas are produced by carburizing and the soft areas are allowed to remain in their original condition by protection from the environment during carburization.

A stop-off compound or a metal spray may be used for protection purposes, but copper plating is preferable as it produces a continuous protective coating.

Any metal that is impervious to carbon monoxide may be used to protect the desired area. Copper is used for such a protective metal because of its high melting point and because copper plating is inexpensive.

For a coating to be successful, it must be thick enough to exclude the carbon monoxide and also it must be free from breaks. If there are pores or cracks in the copper plate, hard spots will form during carburization that will interfere with machining operations or may even cause failure in service.[1]

A copper-plate thickness of less than 0.1 mil has been reported to protect steel, with a good surface finish, against a carburizing pack at temperatures of 1700 to 1740°F.[2] However, thicknesses of 0.5 to 5.0 mils are recommended.[3]

The thickness of copper required may vary with the type of copper bath employed. It has been reported that acid copper must be thicker than cyanide copper for the same degree of protection, but that there is little advantage in using bright copper.[4] The porosity of the deposit is dependent on the type of plating bath for a given steel and the same copper thickness. However, this is not as great a variable as the condition of the steel.

The condition of the steel is very important. It is well known that the porosity of a deposited metal depends greatly on the condition of the surface of the basis metal. If the steel is smooth, a thin coating may be pore-free; but if the steel is roughly machined or is a rough casting, a much thicker deposit will be required.

If the steel is rough, shot blasting may reduce the microscopic roughness and, consequently, reduce the thickness of the copper coating required for adequate protection.

It is good practice to examine the steel for faults before plating and then to examine sections of a finished part after carburizing. Examination of the steel may reveal gross defects such as occlusions in the surface, burrs from machining or cracks or pores in the metal. Any of these faults will be a source of porosity and thus of hard spots. Examination of a section after carburizing may reveal hard spots that can then be traced to a typical fault. Examination after carburizing may also reveal troubles in the plating practice such as failure to obtain bond because of imperfect cleaning, or incomplete removal of scale in the pickling step.

Cracks

If the steel is cracked or porous, troubles may arise during plating, and hard spots may result from imperfect plating. The plated deposit may bridge small cracks, but if it does, a blister may develop at such a point on subsequent heating. If the crack is large, the plating may not bridge the crack, resulting in local failure.

Cracks may be found by microscopic examination of the surface of the steel. If they are not found prior to plating, they may be revealed after plating through spots on the deposit.

Poor Bond

If the cleaning or pickling operations are inadequate, bond of the deposit to the basis metal may not be obtained. Spotty bond will be indicated by blisters on heating and total lack of bond will be revealed by loosening of the entire deposit.

Scale

If the original steel contains surface scale and this is not removed by pickling, then the deposit will not cover at these points and areas for development of hard spots will be present. Any nonconducting inclusion, such as slag or sand, will react in a similar manner.

SLIVERS OR BURRS

Gross irregularities in the steel, such as slivers from a grinding operation or burrs, will not cover completely on copper plating and consequently will result in an imperfect product.

EXAMINATION OF SECTIONS

Microscopic examination of a section after carburizing may reveal the source of the many troubles that lead to hard spots. If the deposit is blistered, sections should be taken through the blister. A crack is occasionally found beneath the blister, but if the blister is large, it is more likely that the steel was not clean prior to plating.

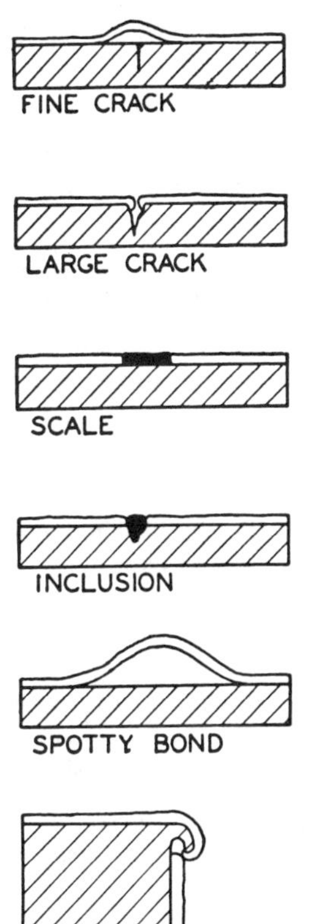

Fig. 27 Causes of poor carburizing results.

Hard spots may be found at a break in the coating because of small particles of nonconducting material. On close examination, it can be usually determined if this is scale, slag or sand.

At other hard spots, slivers of steel from poor grinding or burrs may be detected.

Fig. 27 indicates diagrammatically the appearance of sections from various faults as found by examination after carburizing.

Faults revealed by examination of sections after carburization will suggest steps to avoid further trouble. If the trouble is due to faulty steel, there is little that can be done except to obtain good steel at the start. If the steel is good but excessively rough, mechanical finishing steps will reduce the amount of copper plate required. However, machining operations are expensive, so that it is usually more economical to deposit a heavier copper plate to compensate for the rough surface. Extreme irregularities such as burrs, slivers or undercut machining marks must be removed by shot blasting, tumbling or some other means.

MASKING

In order to confine the copper plate to the desired areas, some method of masking the steel is required. If the areas to be protected are flat, the pieces may be stacked and a number plated at a time. This stacking procedure can be applied very seldom, because the areas to be protected do not often mate on stacking.

For large production runs it is convenient and rapid to have rubber masks made that can be snapped on. For smaller runs, the pieces can be masked with masking tape, wax,[5] or lacquer.

Cyanide Copper Plate

If a copper deposit up to 2 mils in thickness is sufficient for masking purposes, such a thickness may be satisfactorily obtained in a cyanide copper bath. For thicknesses greater than 2 mils, the deposit from the cyanide bath becomes too rough to be of value.

If a greater thickness is required, a thin cyanide copper plate of 0.01 to 0.05 mils is first applied. Such thicknesses are sufficient to insure bonding of a subsequent acid copper deposit.

Acid Copper Plate

Acid copper is deposited over cyanide copper to obtain heavy copper deposits. Smooth, sound deposits of almost any thickness can be deposited from the acid bath. The acid copper, however, cannot be applied directly to the steel since it will not bond.

REFERENCES

1. A. D. Sayers, *The Monthly Review, American Electroplaters' Soc.*, **33,** 1271 (1946).
2. J. C. McCullough and O. M. Reiff, *Ind. Eng. Chem.*, **16,** 611 (1924).
3. R. M. Burns and A. E. Schuh, *Protective Coatings for Metals*, Reinhold Publishing Corp., New York (1939).
4. M. M. Thompson, *Metal Finishing*, **40,** 579 (1942).
5. C. E. Ernst, *Metal Finishing*, **44,** 155 (1946).

37. PLATING BATH TROUBLES

A plating process to be acceptable must insure continuous quality production by test, by observation, and by analysis.

Experienced platers reduce the essential control methods for their plating baths to a minimum by adjustment of the anode area (to maintain proper metal balance in solution), by periodic additions to the baths, and by observation of the appearance of the finished product. Zinc cyanide baths for example, have been successfully operated for years with the aid of frequent on-the-spot analyses for total cyanide and an occasional complete analysis by an outside laboratory. Through correlation of these tests with production results, the plater adds facts to his experience that enable him to operate at maximum efficiency.

A less experienced plater would require frequent analyses of all the important bath ingredients as well as operation instructions explaining the meaning of the analyses and limits on all variables requiring control. However, by providing analytical facilities and specifying narrow bath limits in the instructions, it is possible for a plating engineer to set up a process that can be operated successfully by men with relatively little training. Although there is considerable difference in the methods of these two kinds of operation, the same quality of plate can be obtained in either case and both processes have proved satisfactory.

Analyses may seldom be eliminated from a process, but in some cases, such as a zinc cyanide bath, a plater may never analyze for free caustic and yet he may control it by adding caustic in a definite ratio to the cyanide added, when the cyanide is controlled by analysis.

In the best-controlled process, inferior plating will occasionally be the result of a variable that is not or cannot be controlled or foreseen.

A condition such as a porous basis metal may give rise to a blistered plate. Partially decomposed cooling oil, due to a change in rolling practice at the steel mill, may produce a surface that cannot be cleaned by established practice. Poor plate adhesion from freshly prepared baths may be the result of impurities accumulated in the chemical processing of the plating salts. A rough plate may be caused by suspended solids carried over from the anode sludge, from dirt in the air, or from precipitation of compounds in the bath. The deposit may blister or fail to cover due to improper or inadequate cleaning.

The part to be plated may not be covered completely due to improper anode spacing or arrangement or to poor rack design. The plate may suddenly develop spots that can be due to such a variety of reasons that they may only be corrected through specific knowledge of previous troubles of the same sort or by extensive experimentation.

These common troubles are not readily measurable but are generally detected and controlled by inspection of the finished product.

In case of unexpected troubles, a check should first be made of all controllable limits. If all limits are under control, a check may be made on the cleaning line, as a variety of troubles are caused by poor cleaning. A variable that does not show up in analysis, such as excessive oil in the cleaner, may easily be the offender. It is usually possible to by-pass the cleaner tank by hand cleaning a few pieces with solvent, followed by a mild abrasive paste, and confirm or eliminate trouble at this point or in the pickle by a similar procedure. If the plate is rough, filtration is in order and a rapid check may be made on a small portion of the bath on a laboratory scale before and after filtering.

If impurities in the bath are suspected, a purification step should be tried. Low current density electrolysis, such as is commonly used for nickel baths, may remove metal impurities. Other chemical purification steps can be taken depending on the bath and the impurity suspected. In some cases, it is customary to analyze for and describe limits for tolerance of known impurities, such as iron in a nickel bath.

These general troubles are well known to platers, but the specific troubles found through misfortune and experiment are the offenders that interest us most. The following general troubles do not apply to any particular bath, but are common in many.

General Bath Troubles

Off-Color Deposits

1. *Analysis.* Make a complete analysis of the bath and adjust all ingredients to the proper chemical limits.

2. *Thin Deposits.* Plate a heavy deposit to determine if coverage is adequate to produce the desired color.

3. *Metal Impurities.* Purify by low current density electrolysis or by chemical treatment. Remove the solution from the tank, examine for parts dropped in the tank, for dirt and for electrical faults. Clean the tank and filter the solution into it.

Blistered Deposits or Poor Adhesion

1. *Cleaning and Pickling.* By-pass these steps by using separate tanks or hand cleaning. Examine the part for oil or grease not removed by ordinary cleaning methods.

2. *Fresh Baths.* Very often, deposits from a fresh bath will have poor adhesion, but this trouble will gradually disappear if the bath is electrolyzed for a period of time to remove impurities sometimes present in commercial salts.

3. *Basis Metal.* Examine the basis metal for cracks or holes that may trap solution in the cleaning steps and prevent proper plating.

4. *Strike Baths.* Very often, the addition of a strike step will overcome poor adhesion. A Rochelle copper bath is often used for this purpose.

Rough Deposits

1. *Suspended Material.* Suspended solids in a bath often cause rough deposits. These may be eliminated by filtering.

2. *Poor Anode Corrosion.* If the anode corrodes so that particles of metal form a loose film on the surface of the anode, it is quite common for these particles to be carried to the cathode and cause a rough deposit. Filter the bath and bag the anodes.

3. *Dirt on the Work.* Often particles, such as grinding compound, small rubber or other particles, e.g. those used in previous process steps, are not removed by the ordinary cleaning steps. Hand clean the work with a mild abrasive such as magnesium oxide until no discoloration is obtained after wiping with a clean cloth.

4. *Overplating.* Some baths, such as a copper cyanide bath will invariably produce rough deposits for thicknesses of the order of 0.003 inch or greater.

5. *High Current Density.* Many baths, particularly the acid types, will plate roughly if operated at too high a current density. Cathode agitation permits the use of higher current densities.

6. *Low Metal Content.* Operation at normal current densities with

low metal content will give an effect similar to operation at high current densities with normal metal content. Agitation will help, but addition of metal is more satisfactory.

7. *Agitation.* If agitation is used during plating, such as in the high-efficiency copper cyanide process, frequent or continuous filtration is necessary, since impurities that would normally settle to the bottom of the tank will be suspended by the agitation.

8. *Impurities.* If impurities are suspected, they can only be found by previous experience or investigation. The important fact is that impurities have to enter the bath from some direct contact source—thus, drag-in, chemicals used, water used, anodes, the plating atmosphere, and the work itself may all be suspected. Very often the parts plated, such as zinc-base die castings, may corrode to some extent before they are covered, and eventually cause trouble. Racks that are used for plating in more than one type of bath may easily cause contamination.

Spotty Deposits

Spotty deposits are sometimes caused by poor cleaning and occasionally by pick-up of oil that has accumulated on the surface of the cleaner.

Poor Anode Corrosion

1. *Impurities in Anode.* Specify anode purity.
2. *High Anode-Current Density.* Increase the anode area.
3. *Impurities in the Bath.* Purify the bath and analyze for excessive accumulation of chemicals such as carbonate in alkaline baths.
4. *High Metal Content.* If the metal content is high, the concentration of metal near the anode readily becomes excessive on electrolysis.

Low Free Acid

If the acid is low or a buffering acid is low in a high-pH acid bath, the free acid is readily used up at the anode.

Low Free Cyanide

Has the same effect as low free acid.

Low Cathode Efficiency

1. *Low Metal Content.* If the metal content is low, the ability for hydrogen to evolve is greatly increased.

2. *High Free Cyanide.* Excessive amounts of cyanide will lower the cathode efficiency in many cyanide baths.

3. *Impurities.* Many cases of lowering of cathode efficiency by accumulation of impurities are known, but no general rule can be stated.

In some cases, the accumulation of ammonia in cyanide baths, due to decomposition of the cyanide, will lower the cathode efficiency. This can be remedied temporarily by heating the bath to drive off ammonia.

High Metal Content

Excessive anode area often leads to the build-up of metal in the bath.

Plate Distribution

1. *Recesses.* In general, the acid baths are poor in throwing power and the alkaline baths are good. In case difficulty is experienced in plating in recesses, the location of the anode and the design of the rack should be considered. By allowing the anode to extend into a recess, better covering may be obtained.

2. *Corners and Edges.* A bath that will not plate in recesses will also be a bath that will build up a deposit on corners and edges. This effect may be partially overcome by robbing the high current density areas through the use of auxiliary cathodes near these areas or by the use of nonconducting material between this area and the anode to shadow the area. Each case is a separate problem that must be solved by experience and experiment.

38. CONTINUOUS PLATING

Maximum economy is reached in plating operations by the use of continuous plating machines (usually referred to as automatics) of a semiautomatic or fully automatic nature.

Automatic Plating Machines

In fully automatic plating, the racked unplated work is suspended from hangers at the beginning end of the machine and the finished plated product is removed at the other end. Before entering the machine, the work is usually dirty and unetched, but the operations are so complete that the product at the exit end is fully plated and dry, ready for packing. Other steps, such as bright-dipping and passivating, may be added to the cycle, depending on the requirements of the process and the economy involved. A double rinse or a combination spray and running rinse are sometimes of value. If a high concentration of metal is used in the bath, a drag-out tank may be installed from which metal may be reclaimed, and also strike steps may be added. A flow sheet may read as follows: (1) degrease, (2) electroclean, (3) rinse, (4) acid-dip, (5) rinse, (6) strike, (7) plate, (8) drag-out, (9) rinse, (10) dry.

Such a process would be typical for plating steel with a high-efficiency copper bath, or for plating steel- or copper-base alloys with silver.

In plating zinc-base die castings with copper, nickel, and chromium, a semiautomatic machine may be used, since it is often necessary to buff the pieces either after copper plating or after nickel plating. Thus, the process has to be divided into two parts, each using an automatic machine. One automatic would deliver preplated pieces to the buffing department, which in turn would send them to a second automatic plating machine for finish-plating and inspection.

When plating is put into automatic operation, improvements in the process and improvements in control become attractive. The quanitity of metal consumed per day can be closely observed and the metal going to drag-out and recovery can be easily checked on a large scale. With accumulation of data, improvements can be made in rinsing and drying operations and etching may be changed from chemical to electrolytic etch for closer control.

Automatic plating is applied on an even larger scale where metal is plated as strip or wire before fabrication. The large number of patents that have been issued on various improvements of such machines indicates the detail in which such processes have been studied. Electrotinning is a good example of such a large-scale process.

Electrotinning

Electrotinning replaced hot-dip tinning in many plants because of a demand for thinner tin deposits. This was essential on account of a tin shortage and because the hot-dip process was limited to a minimum thickness of 1¼ pounds per base box (0.075 mil). The electrotinning process has the advantages of producing, at high rates of speed, uniformly distributed deposits of any desired thickness. A thickness of ½ pound of tin per base box was used on a great deal of tinned sheet for the canning industry. This thickness of tin (0.03 mil) does not afford the same protection as 1¼ pounds per base box obtained by the hot dip process, but it was made to serve the purpose by enameling the cans.[1] The enameling operation is expensive, therefore, it is assumed that heavier tin coatings will be used for protection against the corrosive action on the can of foods such as spinach, peas, and tomato juice and that the thinner tin deposits will be used to fabricate containers for less corrosive conditions such as containers for coffee, oil, and paint.

An electrotinning pilot plant was run as early as 1934 and full-scale production was introduced by 1936 by the Carnegie Illinois Steel Corporation.[2] This process, known as the "Ferrostan" process, used an acid tin bath. The plating line was 230 feet long and could be operated at a strip speed of 1000 feet per minute by 4 to 6 operators to produce 1 million base boxes per year (one base box=218 square feet).

The basic steps in this process are: (1) electrocleaning, (2) rinsing, (3) electropickling and chemical pickling, (4) washing and scrubbing, (5) plating, (6) drying, (7) brightening.

The development of such a process required the skill of metallurgists and mechanical engineers for guiding the strip, of electroplaters for the chemistry of the plating, and of electrical engineers for many unique electrical controls.

At the beginning end of all of the electrolytic tinning lines, two coil holders, cropping shears, and a welding unit are arranged. As the steel on one coil holder is being run through the machine, another coil is brought into position and made ready to be welded to the one being plated. In order to provide sufficient time for the welding operation and also to eliminate the necessity of slowing down the line, a looping pit or multiplestrand looper is incorporated in the line. The slack provided by the looper is sufficient for the welding operation, which requires only 4 to 5 seconds and permits continuous high-speed operation on the line.

In high-speed lines of this type, the design of the pinch rolls and tension bridles is important to maintain the proper pull and tension on the strip. Automatic connections are made, including automatic changes in current density to correct for changes in strip speed.

After the preparatory operations are completed, the strip is ready for plating. The plating is carried out either by an acid or by an alkaline process and with the strip traveling either vertically or horizontally through the electrolyte. Each of the processes and methods has its own advantages and disadvantages, but it is not within the scope of this book to go into details of the subject.

In the plating tanks, the solution is sometimes pumped against the strip. This movement of solution and the speed of the strip make it possible to deposit the tin at high current densities and, consequently, at a high plating rate. In fact, such high amounts of current are used in the conductor rolls in some installations that the rolls must be water cooled and in addition special precautions must be taken to prevent arcing which causes defects in the strip.

The high strip speed employed by most lines is accompanied by high-drag-out rates, but efficient drag-out recovery systems reduce the metal loss to a minimum.

After leaving the plating unit, the strip is washed and dried and then usually given a brightening treatment. The brightening step is carried out either by scratch brushing with nickel-silver brushes or by actually melting and flowing the tin over the surface of the steel. At present, practically all of the tinned strip is brightened by melting. The strip is heated to a temperature slightly over 450°F by gas flame, by immersing in hot oil, or by electrical means. Two electrical heating methods have met with favor: (1) conduction, i.e. by passing an electrical current through the strip by contact rolls or (2) induction, i.e. by passing the strip through a high-frequency field of magnetic flux.

After being flow-brightened, the strip is usually given a chromate treatment after which it is either recoiled or cut to size by a flying shear and then automatically separated into prime stock and rejects.

Continuous plating, in some cases using the same equipment, has been applied to electrogalvanizing[3] and plating of steel wire.[4]

REFERENCES

1. K. W. Brighton, *Trans. Electrochem. Soc.*, **84,** 227 (1943).
2. C. Frenkel, *J. Electrodepositors' Tech. Soc.*, **21,** 129 (1946).
3. E. H. Lyons, *Trans. Electrochem. Soc.*, **84,** 279 (1943).
4. J. H. Conolly and R. Rimbach, *Trans. Electrochem. Soc.*, **84,** 293 (1943).

39. PLATING ON PLASTICS

It is possible to plate in some fashion on virtually any type of solid, whether it be metal, ceramic, rubber, leather, wood, glass, or plastic. And there are many methods, other than electroplating, such as vacuum metallizing, silvering, plating over conductive paint, and metal spraying, that have been used to metallize solids.

A general method of plating on nonconductors that consists of variations of the following steps has been developed over a considerable span of time: (1) clean, (2) etch or roughen, (3) sensitize with stannous chloride solution, (4) rinse, (5) deposit a very thin film of a noble metal by chemical reduction, (6) rinse, (7) plate.

This method has been applied where a low level of adhesion could be tolerated.

A somewhat more elaborate process[1] has been developed that has gained wide acceptance because good adhesion to plastic has been attained. The process uses proprietary solutions and only applies to specific plastics. It does apply tosome grades of high-impact ABS and other plastics. The process has gained wide acceptance in a short period of time because of two factors: The plastics to which it applies have good structural properties and the adhesion strength of the plating is good (in fact it is believed that there is an actual bond of the plating to the plastic substrate).

The complete process consists of the following steps: (1) alkaline clean, (2) rinse, (3) neutralize, (4) rinse, (5) etch, (6) rinse, (7) sensitize, (8) rinse, (9) activate, (10) rinse, (11) electroless copper plate, (12) rinse, (13) copper plate, (14) rinse, (15) plate.

The process attracted much interest and when it was realized that for the first time copper-nickel-chromium plating could be commercially applied to plastic parts, it was adopted for large scale application to automotive trim. High production plating of other types of deposits has also been accomplished[2].

This process is elaborate, but elaborate processes are not new to the electroplating industry. Each step is important and has a bearing on the success of the process.

Plastics

A type and grade of plastic must be chosen that is platable. Platable grades are known by suppliers of the proprietary chemicals and by suppliers of acrylonitrile-butadiene-styrene (ABS) and other plastics. Parts must be moded to a design that is platable. Sharp corners, deep recesses, and blind holes should be avoided. The part should also be smooth and free of ragged or uneven surfaces due to molding. These irregularities are magnified by the overlay of a bright reflective metal. If mold release chemicals are used they must be removable by available cleaning methods.

Cleaning

Cleaning can be done in a typical medium duty non-silicated alkaline cleaner.

Rinsing

Double rinsing is generally recommended.

Neutralizing

Neutralizing in a dilute acid dip is frequently recommended (although this step is not always used).

Etching

Etching is done in a chromic-sulfuric acid etch. This step should produce a fine etch on the surface that will condition the surface for subsequent processing and bonding of the electroless copper plate.

Sensitizing

This step consists of absorbing a substance on the surface that will reduce the noble metal salt in the following activation solution. A dilute acidified stannous chloride solution is usually employed for this purpose.

Activation

In this step a reducible noble metal salt reacts with the stannous chloride to deposit neuclei of metal on the surface. The activated surface will function as a catalyst to promote plating of electroless copper (in the following step). A very dilute acidified palladium chloride solution will perform this function.

Electroless Plating (Copper)

The part is now exposed to an electroless copper solution that contains a copper salt, complexing agents, a buffer and a reducing agent.[3] The part is plated by electroless action for 15 to 30 minutes at a rate of 0.05 to 0.1 mil/hr. This covers the surface with a film sufficiently conductive to receive electrolytic plating.

Copper Plating

Copper is electrolytically applied from an acid copper bath. Initial deposition should be done at a low current to avoid damage to the thin electroless film. As the film thickens the current can be increased until plating can proceed under conventional conditions. A copper film of 1.0 to 1.5 mils is typical. A number of proprietary acid copper baths are available.

Plate

Any metal can be applied over the copper electroplate. Usually the copper is sufficiently bright but it can be lightly buffed if desired. Nickel and chromium, nickel and gold, or other metals can be applied over the copper.

This process has been successful when proper attention was given to solution control and to good rinsing for drag-over control. General techniques are quite similar to conventional plating except that more attention to rack design is required. A greater number of contact points are needed to feed the current into the part without damage when the films are thin.

REFERENCES

1. E. B. Saubestre, L. J. Durney and E. B. Washburn, *Metal Finishing*, **62,** 52 (1964).
2. E. A. Blount, *Products Finishing* p. 52 Nov. 1965.
3. U.S. Patent 2,874,072.

40. PREPARATION OF METALS FOR PAINTING

Plating and painting provide two general methods of coating metals. The two methods generally are not competitive because of the substantial differences in properties of the coating and the costs of application. Thus the two remain distinctive and are applied in seperate finishing shops. On the other hand the two methods employ common preparatory techniques such as abrasive cleaning, degreasing, chemical cleaning, and pickling. The plating shop has the environment, equipment, and ability to apply chromate, phosphate, and anodic coatings that are used as a paint base: However, the requirements for the treatments are different when applied as a base for paint and when applied as a final protective coating. This is not always appreciated and in fact the finishing shop often does not know why a chemical conversion coating is being applied.

Because of common problems and common facilities it is inevitable that the problems of the painter enter the plating shop.

The life of a coating depends on the type of paint used and the method of preparation. It is not possible to say which is more important but the more severe the environment the more essential the protective quality of the paint and the more elaborate the required method of preparation. The objective is to prepare and coat the surface sufficiently to attain the expected life.

Steel will stain and rust even indoors. Such steel (painted merely to improve appearance) can be protected with a single coat of paint or primer. In sheltered but unheated environments it is usually required to prime and then paint the steel. In outdoor environments it is reconmended to chemically prepare or treat the steel, prime, and then paint. In mild outdoor environments alkyd paints have been found satisfactory but in more severe outdoor environments more protective paints, particularly phenolics, have been required, although the preparation was the same. Even in more severe, chemical, environments, such as around a plating shop, the treatment is the same except that special chemical resistant paints are required. When requirements are exacting then all basic steps should be specified including the cleaning, treating, priming, and painting.

In broad terms, surface preparation consists of all steps prior to the application of the paint or topcoat. This will include cleaning, chemical treatments, and priming, as required.

The demands for cleaning vary greatly with the condition of the surface as well as the expected service life. Painting for the least demanding condition can be a one step process where the painting step, particularly when brushed on, also performs a cleaning function. Cleaning, prior to painting, may require removal of: (a) nothing; (b) loose corrosion products consisting of rust or scale; (c) organic materials consisting of oil, grease, or old paint; (d) removal of all contaminants present down to the bare or "white" metal; (e) removal of contaminants plus treatment to chemically deoxidise the surface.

A prime coat is one that extends the life of the coating by increasing adhesion and increasing the resistance to corrosive attack where corrosive attack may be either by moisture permeability or by faults in the coating.

Chemical treatments can be a part of the cleaning preparation and also a part of the priming preparation. Pickling and deoxidizing aid the removal of contaminants. Chemical conversion coatings may function as primers.

Chemical Methods

If a single coat of paint or primer is to be applied to a metal; preparation by wiping with a cloth may be sufficient. A prime coat may be sufficiently adherent because it is thin. Multiple coatings will adhere to dry, inert metal if it is sufficiently rough or properly primed. However, oil, grease, dirt, soil, and salts (due to welding, heat treating, or soldering) must be removed. When tight rust and scale can be tolerated the cleaning may be restricted to solvent, alkaline, emulsion, or steam cleaning. If the metal has a good machined or cold rolled surface, hand or solvent wiping is often satisfactory. Small parts can be cleaned by immersion or agitation in solvent and drying by tumbling or passing through sawdust or corncob meal.

Parts are often cleaned in hot alkaline cleaners to remove emulsifiable and saponifiable oils and fats. These metal cleaners contain alkalis, soaps, and wetting agents and cleaning is generally done to a water-break-free condition to assure good removal of oily substances.

Alkaline residues are known to cause difficulty with paint adhesion. After alkaline cleaning the surface must be rinsed to a non-alkaline condition. Rinsing to neutrality is satisfactory but frequently a warm acid dip is used following the rinse to leave the surface in a slightly acid condition. A dilute phosphoric or chromic acid rinse at pH 4.5 is satisfactory.

Emulsion cleaners with hydrocarbons plus emulsifying agents of 1 to 5% by volume in water are used as cleaners by immersion, spray, or wiping. After cold rinsing, hot rinsing, and drying, the parts may be painted. This cleaning method leaves a slight film on the work that is considered acceptable to paint.

On large structures preparation is restricted to hand methods and mechanical portable tools. Loose rust and scale is removed by chipping, scraping, sanding, and wire brushing either with hand tools or power tools. If it is necessary or desirable to remove all rust and scale, this is done by abrasive blasting.

When parts are small an abrasive method can be used that will remove all rust, grit, mill scale, and foreign matter. Methods are available to produce light matte finishes or to do very heavy cleaning, depending on the need. Cleaning can be done with fine or coarse abrasives, steel grit, or shot. Methods may be dry or wet (vapor blast). Barrel finishing is also used to cut, deburr, and descale small parts. The barrels are loaded with burnishing agents such as metal balls and shapes, abrasives, soaps, and cleaners. This type of mechanical cleaning can be done in closed horizontal barrels or open inclined barrels. Heavy and delicate parts can be fixtured to prevent damage.

Abrasive cleaning that leaves only a tight scale is sometimes acceptable and allowed (by specification). A good wetting primer that will cover the scale is then essential prior to topcoating.

Heavy steel is sometimes allowed to weather to aid removal of mill scale by rusting. This loosens the scale, which is then more easily removed by blast cleaning. Artificial weathering can be induced by dipping parts in 50% by volume hydrochloric acid and then allowing to dry and rust. This procedure has also been found to be useful as a pre-pickle treatment.

Flame cleaning plus wire brushing will remove rust, loose scale, and some tight scale from steel. When steel is outdoors and below 40°F it is often heated with a torch to remove thin condensed moisture films.

Complete removal of scale is often accomplished by pickling or by mechanical cleaning supplemented with pickling. When scale is heavy, inhibited pickles are used. Otherwise, conventional pickling practices are used except that pickling times may be longer deliberately to promote etching, which is beneficial to paint bonding. Common pickles include sulfuric, hydrochloric, and phosphoric acids. After pickling the parts are rinsed, dipped in a dilute chromic or phosphoric acid dip, and then dried.

Priming

A prime coat must bond to the metal, to the natural oxide, or to the scale on the metal. It also protects the substrate and provides a coating to which the topcoat will adhere. Some paint systems use several primers, the first coat acting as a bonding coat and the second as an inhibitive coating over which the topcoat is then applied.

A simple primer is one that will adhere and act as a moisture barrier. This is done with a wetting oil primer consisting of linseed oil plus penetrating oils and driers that oxidize to form a water insoluble protective film.

Inhibitive primers consist of a chemically protective pigment dispersed in a suitable vehicle. Zinc chromate pigment is a complex chromate composition that will release chromate ions to water, that permeates the coating, and thus provides an inhibitor that will passivate the metal. Red lead is a basic lead oxide that will react with compounds like linseed oil to form lead soaps. The film is strenghened by these soaps and a tough moisture resistant film is formed. Some moisture will penetrate but the neutralizing and oxidizing characteristics of the pigment will inhibit corrosion. Other inhibiting pigments and mixtures of pigments are used as primers.

A treatment known as "wash" priming or "etch" priming employs protective pigments and phosphoric acid. The primer consists of polyvinyl butyral pigmented with zinc chromate.[1] It is a two part system —one part containing the resin and pigment and the other containing phosphoric acid water and isopropyl alcohol. The two parts are mixed just prior to use. The phosphoric acid reacts with the metal to form an adherent phosphate film which is bonded to the organic film. Wash primer is affective over steel, stainless steel, zinc, cadmium, tin, aluminum, and magnesium. Topcoats of vinyl, phenolics, nitrocellulose, alkyds, and oil paints may be used over the primer.

Protection in severe environments is attained by the use of wash primer, zinc chromate primer over the wash primer, and then as many coats of a suitable paint as are required to provide the desired protection.

Phosphate and chromate chemical conversion coatings function as primers. These treatments form inhibitive coatings that provide a good base for paint. Other primers such as zinc chromate can be applied over these chemical conversion coatings to increase the protective capacity of the under coats but wash primer generally is not applied over these coatings.

Metallic pigments, particularly zinc and aluminum, are used to chemically or electrochemically protect steel.

Zinc and cadmium electroplates might be regarded as prime coats for steel. Such coatings are expensive but when primed and topcoated they will, in some applications, provide the most economical protection. When these electroplated coatings are to be painted they are best phosphated as a step in the coating process.

When conversion coatings or anodic coatings are used as a paint base it has been found that they are best kept thin as compared to the same coatings used for protection of unpainted surfaces. Adhesion is better when the coatings are thin. Anodic coatings will provide a good base when they are thin but do not have to be thin. It is more important not to seal if they are to be painted.

Steel Preparation

The best surface for painting steel is one that is rough and completely free of rust, scale, grease, and dirt, although, as pointed out, this degree of cleanliness is not always necessary. The steel can be cleaned, shot peened or abrasive blasted, and perhaps pickled to produce the best surface. After cleaning to this degree the steel should be primed within not more than 24 hours to avoid the formation of soft rust. A prime coat will protect the steel and in fact a finishing shop is often asked to clean, pickle, and prime steel that can then be held in a protected state. If phosphating or wash priming is used as a prime coat, mechanical roughening can be avoided. If the steel is to be held for some time in the primed condition a stable primer like red lead is preferred. Zinc chromate primer may be applied over a phosphate coating or a wash primer. Sand blasted steel is generally primed with red lead or zinc chromate.

Aluminum Preparation

Anodizing, phosphating, chromating, or other conversion coatings will prepare aluminum for painting. The preparation for these treatments is the same as when they are used for protective coatings. However, post treatment usually must be different. The anodic coating is best not sealed. If a part of the surface is to be left unpainted it can be sealed after painting, providing that the paint is a type that will withstand the hot water seal treatment. Warm post-rinse treatments that are beneficial for protective conversion coatings are not necessarily best for parts to be painted. Hotter rinses that leach the film may be advantageous, depending on the type of conversion coating. Phosphate coatings are typically followed by an acidified rinse in dilute chromic or phosphoric acid. A hot alkaline chromate solution can be used to produce a thin conversion coating on aluminum that will provide a base for paint.

Wash primers can be applied to aluminum. Zinc chromate primers can be applied over the wash primer or over anodized or conversion coated surfaces.

Zinc and Cadmium

Zinc or cadmium should be phosphated, chromated, or wash primed. Zinc chromate can then be applied and a further top coat applied if desired.

REFERENCES

1. U.S. Patent 2,525,107.

41. ANALYTICAL METHODS FOR PLATING BATHS

Analytical methods are the foundation of plating-bath control. Although it is possible to go far in controlling many baths without resorting to analytical methods, it is not necessarily desirable to eliminate analyses. By the use of analytical control it is possible to hold a bath in limits, whereas if control were maintained purely by physical and plating tests, the bath would gradually drift away from the original limits. Eventually a point would be reached where radical changes in the bath would be required to restore it to the original limits. This usually results in temporary shutdown of the tank and loss of time and production.

Analyses are not made to get out of trouble but rather to stay out of trouble. By frequent analyses and continuous adjustment of chemicals, baths have been operated for years without trouble.

It is worthwhile both to review analytical methods recommended for plating-bath control and to take an occasional look at the standard analytical books. Plating methods are rapid and relatively inaccurate compared to standard routine analytical methods. This relative inaccuracy is allowable for the wide limits used for plating control, but it should be realize that if accuracy is required a better method is preferable. For instance, such accuracy is required to check the purity of plating salts.

Analytical methods are dependent on the purpose of the analysis, so that it is not possible to use a single method that will be satisfactory for all applications of a substance. A particular method is selected according to the relative importance of the factors—speed, cost, and accuracy. Each of these factors assumes a different importance in each application. A method used to control the metal in a flash bath may not be satisfactory to control the metal in a bath used to produce heavy deposits. Likewise, methods used to control a bright plating bath may not be required for a bath used to produce deposits for resistance to corrosion.

Many times it is desirable to have more than one method for the same substance. Two different methods may be used to check the results of one against the other or an analyst may prefer a certain analysis. Of the many methods available, the following are examples of various ways of determining copper in a copper cyanide bath:

1. Destroy the cyanides and plate the copper from an acid solution on a weighted platinum cathode.
2. Destroy the cyanides with acid and titrate with sodium thiosulfate.
3. Determine colorimetrically as the blue complex formed with ammonia or by the color produced from copper chloride in a strong hydrochloric acid solution.
4. Determine polarographically by the addition of the proper electrolyte.
5. Precipitate as copper hydroxide and determine the volume of precipitate with a centrifuge.
6. Estimate the copper from the titration of free cyanide and the cathode efficiency.

The selection of one of these methods depends on the accuracy desired and the equipment available. If a centrifuge is available for the determination of sulfate in a chromic acid bath, it may also be used for the determination of copper. If platinum electrodes are available for the electrolytic determination of copper, then these may be also used for the electrolytic determination of cadmium.

Volumetric Methods

The volumetric methods are very popular. They are relatively fast and usually quite accurate. They depend on the quantitative completion of a chemical reaction by addition of one solution to another. The volume of the solution that is added to the sample is accurately measured. The completion of the first reaction is indicated by a secondary reaction which proceeds with a change in color or with the formation of a precipitate. The accuracy of a volumetric method depends on:

1. The completeness of the primary reaction.
2. The sharpness of the end point indicated by the secondary reaction.

3. The concentration of the standard solution (usually expressed as normality). The chemists have determined the best reactions and the most suitable indicators for a particular method, but it is up to the analyst to prepare and test his standard solutions. For plating analyses, most standard solutions may be made by weighing enough chemical to make a solution of the desired strength. However, it should be realized that accurate solutions require standardization. It is not possible to make a solution of hydrochloric acid that will be 0.1000N ±0.0002N unless the solution is carefully standardized and adjusted. However, it is possible to make a silver nitrate solution of this accuracy without standardization, since silver nitrate is a pure salt and may be used as a standard for other slutions.

As an example; silver nitrate may be used to determine the value of a hydrochloric acid solution (by neutralizing a known volume of solution and titrating for chloride). Silver nitrate, however, is limited as a standard, since it must be freshly made. The difference between a fresh solution and an old one is very important. Most standard solutions decompose with age. Solutions must either be made up often, or they must be occasionally standardized. The best standard solutions are those that are stable, for example, a sodium chloride solution. The best methods for standarization are those that use exactly the same steps as the analytical method. The best standard solution to standardize silver nitrate for the determination of chloride in a nickel bath would be a standard nickel bath. In other words, a nickel bath similar to that being analyzed with a known amount of chloride. In fact, the same nickel bath could be made to serve as a standard for sodium hydroxide solution for the determination of boric acid and as a standard for sodium cyanide solution for the determination of nickel.

Gravimetric Methods

Gravimetric methods are not widely used for plating-bath control since they are slower than the other types of methods. The gravimetric methods consist of careful separation of a chemical compound from other substances in the bath. The compound is precipitated, washed, dried, and weighed. These methods require a balance and an oven, depending on the drying step. The gravimetric methods are generally accurate and are often used for check work where a substance is determined by more than one method.

Electrolytic Methods

Electrolytic methods may come under the general classification of gravimetric methods since the metal is weighed. In the electrolytic methods the metal sought is deposited on a platinum cathode. It is then washed, dried, and weighed. These methods are usually faster than the other gravimetric methods and are quite accurate. The high accuracy that can be obtained is indicated by the fact that the electric current is standardized in terms of plating silver from a silver nitrate solution.

Cathode-Efficiency Methods

The amount of metal in a bath can often be estimated by determination of the cathode efficiency. This may be done by plating at a known current density for a definite period of time. The amount of metal on the cathode may be determined by weighing or it may be determined by deplating at low current density and noting the time required. These methods can only be used if the effect of metal concentration, temperature, and concentration of other chemicals on the cathode efficiency are known. The accuracy is only fair, but very rapid methods can be developed.

The Centrifuge

The centrifuge may be used for the rapid determination of an ingredient that can be precipitated.[1] Carbonate can be determined by the addition of barium chloride. Sulfate can be determined in a chromic acid bath. Free cyanide can be determined in a copper cyanide bath by the addition of copper sulfate solution. Many of the metals can be determined by centrifuging if they can be precipitated as a sulfide, sulfate, or other insoluble compound.

The pH Meter

The pH meter is valuable for the direct determination of pH and as a means of recognizing the end point where a titration is made to a definite pH.

The Colorimeter

Colorimetric methods are rapid and reasonably accurate where they can be applied. Specific methods must be worked out so that the color is measured at the proper wave length. Nickel can be determined colorimetrically in a few minutes. It is not necessary that the solution is colored if it will absorb light rays, as shown by a method used for the determination of nitrate in silver baths.[2]

Colorimetric methods are often valuable to determine impurities in plating baths with a view to their sensitivity to small amounts of metal.[3]

The Polarograph

The polarograph is an instrument that measures the voltage and amperage characteristics of a dropping mercury electrode. It has been applied to the rapid simultaneous determination of copper and zinc in brass deposits.[4] It is obvious that this instrument can be used for the determination of the metal content of most of the plating baths. However, as with the colorimeter, much experimental work is required before it can be determined how the instrument will compete with available methods. It is suitable for the analysis of alloy baths and deposits.

The polarograph, like the colorimeter, can be used to determine small amounts of impurities in plating baths.[5]

Methods Still Needed

There are many ingredients that cannot be determined by known methods; most of these are addition agents. These complex organic substances do not lend themselves to ready analysis; if they are colloidal, the analysis may be of no value. One may analyze for gelatin in an acid bath by determination of the total nitrogen content, but the analysis will not tell how the bath will operate. Only part of the gelatin participates actively in the process, the rest may only be a harmful residue. The only answer to this problem is to control the bath by regular addition, by observation of the plate, or by plating-cell control.

Specific Analytical Methods

The following methods are specific for the analysis of the given substances in a plating bath. They should be used as a reference list. Alternatives are given and new methods of analysis are suggested so that one may select one's own set of analytical procedures. It is well for the plater to do this since each application may call for a different method, depending on the equipment available and the accuracy required. The list indicates that many methods can be developed that are not specific plating methods. New methods may be developed from standard analytical texts or from a knowledge of general methods such as the application of a pH meter to analytical work.

Determination of Acids

Boric Acid in a Nickel Bath. Pipette a 5 ml sample into a 250 ml Erlenmeyer flask. Add 25 ml glycerin. Add 0.5 ml bromcresol purple indicator solution. Titrate with N/10 sodium hydroxide until the solution changes to dark green and then to a blue purple.

$$\text{ml NaOH} \times 0.166 = \text{oz/gal } H_3BO_3$$

It is well to standardize the sodium hydroxide with a nickel bath containing a known amount of boric acid. The same size of sample and exactly the same procedure should be used as in the analysis. This will enable the analyst to recognize the proper color change and all errors in the method will be alike during the analysis and the standardization, thus canceling the errors.

Undersaturation Method. Add 3 to 6 g boric acid to 250 ml of the bath. Add a few drops of wetting agent (e.g. Tergitol). Agitate for 1 hour. Filter on a small Buchner funnel. Wash with 10 ml acetone and allow to dry. Pull dry for 15 minutes. The residue contains 20% liquid.

In some plating baths, such as the fluoborate bath, boric acid is very difficult to determine. This method, which may be called a measurement of undersaturation may be used in such a case. If it is used, it is best to redefine the limits for boric acid in terms of undersaturation. For instance, if the limits call for 25 to 35 g/l and a saturated bath would contain 45 g/l, then the undersaturation limits are 10 to 20 g/l. The residue weighed is a direct measure of undersaturation where the residue holds 20% liquid. A correction should be made for dry weight or the residue should be dried.

Chromic Acid. Pipette 10 ml of the sample into a 500 ml volumetric flask. Dilute to the mark and mix thoroughly. Pipette 10 ml from the volumetric flask into a 250 ml Erlenmeyer flask. Add 100 ml distilled water. Add 2 g ammonium bifluoride. Add 15 ml concentrated hydrochloric acid solution and 10 ml of a solution containing 100 g/l potassium iodide and 1 g/l potassium hydroxide. Titrate with N/10 sodium thiosulfate until the brown color fades to straw. Add 5 ml starch solution and titrate until the blue color disappears.

$$\text{ml } Na_2S_2O_3 \times 2.234 = \text{oz/gal } Cr \cdot O_3$$

Fluoboric Acid. Fluoboric acid may be determined by conductivity if correction is made for the metal content of the bath.

The correction for the metal content may be made by a gravity measurement.

Free Fluoboric Acid in a Lead Fluoborate Bath. Pipette a 10 ml sample into a 400 ml beaker. Dilute to 250 ml with distilled water. Titrate with N/1 sodium hydroxide until the first permanent white cloudiness appears.

In most analytical procedures, recommended volumes are only approximate, but in this procedure the recommended volume must be exactly the same for each determination.

$$\text{ml NaOH} \times 1.18 = \text{oz/gal } HBF_4$$

Free Acid. Pipette a 25 ml sample into a 250 ml Erlenmeyer flask. Add 125 ml distilled water and 5 drops of methyl orange indicator. Titrate with N/1 sodium hydroxide.

$$\text{ml NaOH} \times 0.196 = \text{oz/gal HCl}$$

$$\text{ml NaOH} \times 0.263 = \text{oz/gal } H_2SO_4$$

This is a general method for free acid in the presence of any metal that does not precipitate before the methyl orange end point is reached.

Hydrochloric Acid and Iron. Hydrochloric acid can be determined in the presence of iron by the conductivity method, but a correction must be made for the iron. This correction can be made by a gravity measurement. By properly setting up original standards, acid and iron can thus be run by the measurement of conductivity and gravity.

Sulfuric Acid in Acid Copper. Pipette a 10 ml sample into a 250 ml beaker. Add 100 ml distilled water and 5 drops of methyl orange indicator. Titrate with N/1 sodium hydroxide until the color of the solution changes from violet to pale green.

$$\text{ml NaOH} \times 0.657 = \text{oz/gal } H_2SO_4$$

Ammonia

Cyanide Solutions. Pipette a 20 ml sample into a 1 l distilling flask. Place 75 ml 10% sodium hydroxide solution in the dropping funnel. Pipette 25 ml of N/10 hydrochloric acid into the receiver and add 1 drop of methyl red indicator. Slowly run the sodium hydroxide solution into the distilling flask, leaving a small amount in the funnel to act as a seal. Distill off the ammonia and watch that the color of the indicator does not change. Back-titrate the excess hydrochloric acid with N/10 sodium hydroxide. Run a blank to determine the exact milliliters of N/10 sodium hydroxide required to titrate 10 ml of N/10 hydrochloric acid.

$$(\text{ml blank—ml titration}) \times 0.0234 = \text{oz/gal } NH_4OH$$

Brass Bath. Monaweck[6] reported that the precipitates of $NaCu(CN)_2$ and CuCN decomposed during distillation to convert CN to NH_3. To eliminate these precipitates, he suggests the following method: Pipette a 10 cc sample of the brass solution into an Erlenmeyer flask. Add 25 cc distilled water. Add 25 cc of cold 1 to $4H_2SO_4$. Filter off precipitate. Wash precipitate with 50 cc water. Evaporate filtrate until SO_3 fumes begin to appear. Cool and dilute to 250 cc. Add a few drops of phenolphthalein. Add $Ca(OH)_2$ solid, until the slurry turns pink. Distill for ½ hour in a Kjeldahl apparatus, collecting the NH_3 distillate in standard acid. Titrate the excess acid.

$$(\text{ml blank—ml titration}) \times 0.0468 = \text{oz/gal } NH_4OH$$

Cadmium

Cyanide Solutions—Alcohol Method. Pipette 5 ml sample into a 25 ml volumetric flask. Make up to the mark with methyl alcohol. Shake and allow the crystals to settle. Filter dry into a beaker. Pipette 10 ml into a 100 ml beaker. Add 10 ml 1 to 2 hydrochloric acid and 40 ml water. Titrate with N/10 potassium ferrocyanide.

$$\text{ml } K_4Fe(CN)_6 \times 0.752 = \text{oz/gal Cd}$$

Cyanide Solutions—Electrolytic Method. Pipette a 5 ml sample into a 400 ml electrolytic beaker. Add 75 ml distilled water. Electrolyze with platinum electrodes at 1.5 amperes for 2 hours, using a weighed cathode. Dry and weigh the cathode.

$$\text{Cd deposited} \times 26.8 = \text{oz/gal}$$

Cyanide Solutions—Volumetric Method. Pipette a 5 ml sample into a 400 ml beaker. Add 50 ml distilled water and 10 ml of 15% sodium sulfide. Warm the solution and filter off the cadmium sulfide precipitate. Wash the precipitate thoroughly with warm distilled water and discard the filtrate. Transfer the paper and precipitate into the original beaker. Add 25 ml concentrated hydrochloric acid and 25 ml distilled water. Heat to boiling to dissolve the precipitate. Dilute to 200 ml with distilled water and add ammonium hydroxide until it turns the red litmus paper blue. Add concentrated hydrochloric acid until the litmus paper turns red again and then add 3 ml in excess. Heat nearly to boiling and titrate with N/10 potassium ferrocyanide using uranium acetate as an outside indicator.

The presence of zinc will cause an error in this method.

$$\text{ml } K_4Fe(CN)_6 \times 0.300 = \text{oz/gal Cd}$$

Carbonate

Cyanide Solutions. Pipette a 10 ml sample into a 150 ml beaker. Add 1 ml concentrated ammonium hydroxide and 100 ml 10% barium chloride solution. Heat to boiling and boil gently for 10 minutes. Filter and wash the precipitate with hot water. Add a few drops of barium chloride solution to the filtrate to determine if the precipitation was complete. Transfer the filter paper and precipitate into the original beaker. Add 50 ml distilled water and a few drops of methyl orange indicator. Titrate with N/1 hydrochloric acid to a permanent pink color.

$$\text{ml HCl} \times 0.71 = \text{oz/gal } Na_2CO_3$$

The accuracy of this method is only fair, which is all that is usually required. Greater accuracy may be obtained by distilling the carbon dioxide with sulfuric acid, collecting it by precipitation as it is bubbled through a barium hydroxide solution, and then treating the precipitate as above. For greater detail on carbonate analyses consult the analytical books.

Cyanide Baths[7]*—Centrifuge Method.* Pipette a 5 ml sample into a 100 ml Goetz centrifuge tube. Add 4 drops of wetting agent and 45 ml distilled water. Shake the tube so that its contents are thoroughly mixed. Place on a water bath above 200°F for 5 minutes. Add 50 ml barium chloride solution (25 g/l) also above 200°F. Shake and then place on a hot water-bath for 10 minutes. Remove from the bath, shake and then spin in centrifuge at 2000 rpm for 5 minutes. Read the height of the precipitate in the tube and derive the concentration of sodium carbonate in ounces per gallon from a prepared standard curve.

Chloride

Pipette a 5 ml sample into a 250 ml Erlenmeyer flask. Add 50 ml distilled water and 1 ml 10% potassium chromate solution. Titrate with standard N/10 silver nitrate solution until a reddish color appears.

$$\text{ml } AgNO_3 \times 0.095 = \text{oz/gal Cl}$$

Chromium

Trivalent Chromium in a Chromic Acid Bath. Pipette a 10 ml sample into a 500 ml volumetric flask. Dilute to the mark and mix thoroughly. Pipette 10 ml from the volumetric flask into a 250 ml Erlenmeyer flask. Add 100 ml distilled water. Add about 0.2 g sodium peroxide. Boil for 20 to 30 minutes. Dilute to 100 to 125 ml and cool. Add 2 g ammonium bifluoride and proceed as for the analysis of chromic acid.

$$\text{ml } Na_2S_2O_3 \text{ used for } Cr^{+++} - \text{ml } Na_2S_2O_3 \text{ used for } CrO_3 \times 1.16 = \text{oz/gal } Cr^{+++}$$

Copper

Electrolysis Method for Cyanide Solutions. Pipette a 20 ml sample into a 200 ml electrolytic beaker. Add 5 ml concentrated nitric acid and 5 ml concentrated sulfuric acid. Boil until dense white fumes of sulfur trioxide are evolved. Cool. Slowly add 100 ml distilled water. Weigh a platinum-gauze cathode and set up the platinum cathode and a platinum anode for electrolysis. Lower the electrodes into the beaker until the solution nearly covers them and start agitation (agitate with a stirrer or with air). Deposit the copper at 4.0 amperes until the blue color disappears and then continue the electrolysis for an additional 15 minutes at 2.0 amperes. Add water so that the level of the solution is raised and an unplated portion of the cathode is exposed to the current. If no further copper deposits, the electrolysis is complete. Without interrupting the current, stop the agitation and slowly raise the electrodes while washing the cathode down with a continuous steam of water to prevent resolution of the copper. Rinse the cathode in alcohol, then in ether, and dry at 50°C. Weigh. From the gain in weight determine the copper content.

$$\text{g Cu} \times 6.7 = \text{oz/gal Cu}$$

Volumetric Method for Acid Solutions. Pipette a 2 ml sample into a 250 ml beaker. Add 25 ml distilled water. Add concentrated ammonium hydroxide until the solution turns dark blue. Boil for 15 minutes. Add 10 ml 30% acetic acid solution. Cool to room temperature. Add 25 ml 20% potassium iodide solution. Swirl the contents to mix. Titrate with N/10 sodium thiosulfate using starch as an indicator.

$$\text{ml } Na_2S_2O_3 \times 0.426 = \text{oz/gal Cu}$$

Volumetric Method for Cyanide Solutions. Pipette a 10 ml sample into a 250 ml Erlenmeyer flask. Add 5 ml concentrated sulfuric acid and 0.5 ml concentrated nitric acid. Boil until dense white sulfur trioxide fumes are evolved. Cool. Add 100 ml distilled water. Add concentrated ammonium hydroxide until the solution turns dark blue. Boil for 15 minutes. Add 10 ml 30% acetic acid solution. Cool to room temperature. Add 25 ml 20% potassium iodide solution. Swirl the contents to mix. Titrate with N/10 sodium thiosulfate solution until the brown color fades to yellow. Add 5 ml 1% starch solution and continue titrating until the blue color disappears and does not reappear for at least 1 minute.

$$\text{ml } Na_2S_2O_3 \times 0.0852 = \text{oz/gal Cu}$$

CYANIDE

Free Cyanide in Copper Cyanide. Pipette a 5 ml sample into a 150 ml beaker. Add 100 ml distilled water. Add a few drops of 10% potassium iodide solution. Titrate with N/10 silver nitrate to a slight turbidity.

$$\text{ml } AgNO_3 \times 0.261 = \text{oz/gal free NaCN}$$

This method is reliable for control of a Rochelle copper bath, but it must be modified to control a high-efficiency copper cyanide bath.

Free Cyanide in a High-Efficiency Copper Bath. Pipette 10 ml of the bath into a dry flask. Add 3 g Rochelle salt and 1 g potassium iodide. Agitate until the salts are dissolved. Titrate with N/10 silver nitrate.

$$\text{ml } AgNO_3 \times 0.130 = \text{oz/gal free NaCN}$$

This method will give approximately the same result as the cold titration for a high-efficiency sodium bath. For the potassium bath, it is believed to be more reliable than the cold titration.

Free Cyanide in High-Efficiency Copper Bath. Place 10 ml of 10% potassium iodide solution and 40 g cracked ice into a 250 ml Erlenmeyer flask. Pipette 10 ml plating solution into the flask and stir until a temperature of 3 to 5°C is reached. Titrate slowly at this temperature with N/10 silver nitrate until a distinct opalescence appears which is permanent for 1 minute.

$$\text{ml } AgNO_3 \times 0.131 = \text{oz/gal free NaCN}$$

Total Sodium Cyanide in Cadmium Cyanide. Pipette 2 ml sample into a 250 ml Erlenmeyer flask. Add 100 ml distilled water, 5 ml 100% potassium iodide, and 15 ml concentrated ammonium hydroxide. Titrate with N/10 silver nitrate to a faint yellow turbidity.

$$\text{ml } AgNO_3 \times 0.65 = \text{oz/gal NaCN}$$

Total Sodium Cyanide in Cadmium Cyanide. Pipette a 1 ml sample into a 250 ml Erlenmeyer flask. Add 100 ml distilled water. Add 5g of a dry powdered mixture of 95% by weight sodium bicarbonate and 5% by weight potassium chromate. Dissolve the salts. Titrate with N/10 silver nitrate containing 10 ml per liter concentrated nitric acid to the first permanent brick-red color.

$$\text{ml } AgNO_3 \times 1.30 = \text{oz/gal NaCN}$$

If iron or ferrocyanide is present, it will introduce an error in this method.

Free Cyanide in Silver and Gold Baths. The general method for total sodium cyanide may be used for the determination of free cyanide in gold and silver baths.

Total Sodium Cyanide. Pipette a 10 ml sample into a 250 ml Erlenmeyer flask. Add 20 ml 10% potassium iodide solution and 100 ml distilled water. Titrate with N/10 silver nitrate to the first turbidity.

$$\text{ml } AgNO_3 \times 0.130 = \text{oz/gal free NaCN}$$

This is the general method for determining sodium cyanide. If sodium cyanide alone is present it will measure the total sodium cyanide present. In a silver bath it will only measure the free cyanide. The method may be altered for each bath, depending on the metal present.

Total Cyanide in Zinc Cyanide Solutions. Pipette a 5 ml sample into a 250 ml Erlenmeyer flask. Add 100 ml distilled water, 5 ml 20% sodium hydroxide solution, and 5 ml 10% potassium iodide solution. Titrate with N/10 silver nitrate to a faint turbidity.

$$\text{ml } AgNO_3 \times 0.261 = \text{oz/gal total NaCN}$$

Gelatin

Take a 25 ml sample. Add 15 ml sulfuric acid, 4 g potassium hydrosulfate, and 0.1 g selenium. Digest for 30 minutes. Dilute to 300 ml. Add 40 ml 40% sodium hydroxide. Distill into 10 ml N/10 hydrochloric acid.

Titrate the remaining hydrochloric acid with N/10 sodium hydroxide. Run a blank titration on 10 ml of N/10 hydrochloric acid.

$$(\text{ml blank} - \text{ml titration}) \times 0.0502 = \text{oz/gal gelatin}$$

Gold

Analysis of Gold and Gold-Alloy Solutions. 1. Pipette 25 ml gold plating solution (or the volume which contains 20 to 50 mg gold) into a 400 ml beaker. This size sample is correct for a 3 to 5 pennyweight per gallon bath.

2. Under a fume hood, add 25 ml concentrated hydrochloric acid, stir and bring to a boil. Add 5 ml concentrated nitric acid and boil for 5 minutes.

3. Remove from flame and add slowly, with agitation, 25 ml of a 5% sodium hypochlorite solution. Then add 35 ml water. Mix thoroughly.

4. Boil gently for 10 minutes.

5. Allow to cool and add, with stirring, 20 ml 20% sodium hydroxide solution. From a pipette or burette, add a saturated solution of

sodium bicarbonate in portions of 5 to 7 ml. After each addition, stir vigorously and test the acidity by touching a piece of red litmus paper with the end of the stirring rod. The drop on the stirring rod should be as small as possible. A faint trace of blue in the litmus paper is taken as the end point for the addition of bicarbonate.

If, after the addition of sodium hydroxide, the sample is alkaline, add 1 ml dilute hydrochloric acid at a time until the sample is acid to litmus. Then add bicarbonate as described above.

6. Add 20 ml 30% potassium iodide solution all at once with agitation. Allow to stand for 2 minutes.

7. Add 2 g solid sodium bicarbonate and titrate with a standard N/100 arsenious oxide solution.

8. When the yellow-red iodide color becomes a pale straw yellow, add 2 ml of a 1% starch solution and continue the titration to the disappearance of the starch-iodine color.

In the presence of blue-colored basic compounds, e.g. copper, it is best to add the starch solution at the beginning.

Cyanide Solution. Pipette a 10 ml sample into a 250 ml beaker. Add 15 ml concentrated hydrochloric acid and evaporate to a syrupy consistency over a water- or steam-bath. Add 150 ml of distilled water. Add 25 ml 20% potassium iodide solution and 2 ml 1% starch solution. Titrate with N/100 sodium thiosulfate to complete disappearance of the blue color. Back titrate with N/100 iodine solution to reappearance of the blue color.

Reagents of N/100 normality are not stable enough to use without standardization. Therefore, these reagents must be standardized with a known amount of gold.

Electrolytic Method. Pipette a 5 ml sample into a 250 ml electrolytic beaker. Add 200 ml distilled water. Add 5 g sodium hydroxide and 5 g sodium cyanide and dissolve. Electrolyze at 4 to 6 volts for about 45 minutes on a weighed platinum cathode. Rinse with distilled water, dry in an oven, cool and weigh.

$$\text{g Au} \times 200 = \text{g/l Au}$$

$$\text{g Au} \times 26.8 = \text{oz/gal Au}$$

Indium

Cyanide Bath. Pipette a 10 ml sample into a 250 ml beaker. Add 10 ml concentrated hydrochloric acid. Add 25 ml distilled water and boil to expel the hydrogen cyanide. Add 5 drops of methyl-orange indicator. Add concentrated ammonium hydroxide slowly until the solution turns yellow. Bring to a boil. Filter and wash. Ignite in a good oxidizing atmosphere at about 800°C. Weigh as In_2O_3.

$$\text{oz } In_2O_3 \times 11.1 = \text{oz/gal In}$$

Iron

Chloride Bath. Pipette a 5 ml sample into a 250 ml Erlenmeyer flask. Add 125 ml distilled water and 15 ml of sulfuric-phosphoric mixture consisting of 150 ml concentrated sulfuric acid and 150 ml concentrated phosphoric acid diluted to 1 l. Add a few drops of diphenylamine solution consisting of 1 g diphenylamine dissolved in 100 ml 60% acetic acid. Titrate with N/2 potassium dichromate until the solution changes from green to violet.

$$\text{ml } K_2Cr_2O_7 \times 0.748 = \text{oz/gal Fe}$$

Sulfate Bath. Pipette a 5 ml sample into a 250 ml Erlenmeyer flask. Add 100 ml distilled water and 5 ml sulfuric acid solution consisting of 1 part sulfuric acid to 1 part water. Titrate with N/2 potassium permanganate to a permanent pink color.

$$\text{ml } KMnO_4 \times 0.748 = \text{oz/gal Fe}$$

Lead

Gravimetric Method. Pipette a 10 ml sample into a 400 ml beaker. Add 100 ml distilled water. Add 10% sulfuric acid until precipitation is complete. Heat almost to boiling and keep hot for 15 minutes. Cool. Filter and wash with cold 1% sulfuric acid. Ignite at 800°C. Weigh as $PbSO_4$.

$$\text{g } PbSO_4 \times 9.1 = \text{oz/gal Pb}$$

Volumetric Method. Pipette a 5 ml sample into a 400 ml beaker. Add 10 ml concentrated nitric acid and 15 ml concentrated sulfuric acid. Evaporate to dense white fumes. Cool. Add 150 ml distilled water and 10 g tartaric acid. Heat to boiling. Cool. Filter off the lead sulfate and wash with cold water. Transfer the paper to the original beaker and add 30 ml saturated ammonium acetate solution. Add 150 ml distilled water. Heat to boiling and titrate the hot solution with standard ammonium molybdate solution using tannic acid as an external indicator. The end point is the first faint yellow color.

$$(NH_4)_2MoO_4 \times 0.134 = \text{oz/gal Pb}$$

Nickel

Pipette a 5 ml sample into a 250 ml Erlenmeyer flask. Add 50 ml distilled water and 0.5 ml 10% potassium iodide solution. Add concentrated ammonium hydroxide until the solution turns a clear blue. Add 0.5 ml N/10 silver nitrate. Titrate with N/2 sodium cyanide until the precipitate which first forms just redissolves.

$$\text{ml NaCN} \times 0.393 = \text{oz/gal Ni}$$

Resorcinol

Dilute a 10 ml sample to 50 ml. Add 2 g 20-mesh zinc and allow to stand for 10 minutes. Add 50 ml water and 5 ml hydrochloric acid and allow to stand 30 minutes longer. Filter and wash. Add potassium bromide and standard potassium bromate solution. Allow to stand for 15 minutes. Add potassium iodide. Allow to stand 15 minutes and titrate with N/10 sodium thiosulfate. Run a blank without the sample and the zinc, but with the timing the same as with the sample, to determine the value of 50 ml potassium bromate.

$$(\text{ml blank} - \text{ml titration}) \times 0.0245 = \text{oz/gal resorcinol}$$

Potassium bromate solution	3 g/l
Potassium bromide solution	250 g/l
Potassium iodide solution	25%

Rochelle Salt

Cyanide Solutions. Pipette a 5 ml sample into a 250 ml volumetric flask. Add 100 ml distilled water and a particle of solid phenolphthalein. Add 1% sulfuric acid solution, drop by drop, until the pink color just disappears. Add exactly 5 ml nitrobenzene. Add 10% silver nitrate solution, drop by drop, until the color of the precipitate changes from white to blue-gray. Shake for ½ minute to coagulate the precipitate. Allow to settle and add 1 drop of silver nitrate to determine if the precipitation is complete. Bring the solution up to the mark, add exactly 5 ml distilled water above the mark and shake thoroughly. Allow to stand until the precipitate has settled. Pipette 50 ml into a 500 ml Erlenmeyer flask. Add 5 ml 20% sulfuric acid, 5 g manganese sulfate, and 100 ml distilled water. Heat to about 70°C. Add slowly and with agitation exactly 20 ml N/10 potassium permanganate solution. Allow the hot solution to stand for 5 minutes. Cool to room temperature under running water. Add 2 g solid potassium iodide and 2 ml 1% starch solution. Shake the flask for a few seconds. Titrate with N/10 sodium thiosulfate until the blue color disappears and does not reappear for at least 1 minute.

The potassium permanganate and sodium thiosulfate should be standardized for this method.

$$6.28(20 \times \text{N of } KMnO_4 - \text{ml } Na_2S_2O_3 \times \text{N of } Na_2S_2O_3) = \text{oz/gal } KNaC_4H_4O_6 \cdot 4H_2O$$

SILVER

Pipette a 5 ml sample into a 250 ml beaker. Add 100 ml distilled water and 10 ml of 15% sodium sulfide solution. Heat to boiling. Filter and wash precipitate with hot water. Transfer the paper and precipitate to the original beaker and add 50 ml of 1-2 nitric acid. Boil to dispel the precipitate. Add 10 ml 10% ferric ammonium sulfate. Titrate with N/10 potassium thiocyanate solution to the appearance of a pink coloration.

$$\text{ml KCNS} \times 0.372 = \text{oz/gal AgCN}$$

Electrometric Determination. This method illustrates how a cell may be set up to titrate to a definite potential. The method is similar to titration to a definite pH with a pH meter except that a system has to be used that will measure the potential of the metallic ion rather than that of the hydrogen ion.

An apparatus was developed so that both free cyanide and silver could be determined from one titration.

Cyanide Strike Bath. Silver can be determined in a silver-strike bath by plating and deplating. This is a rapid method of determining the cathode efficiency which is a measure of the silver concentration. By diluting a silver bath with a sodium cyanide solution this method can be used for the determination of silver in a standard silver cyanide bath.

Cyanide Bath. Pipette 10 ml bath into a 400 ml beaker. Dilute to 100 ml Add nitric acid drop by drop, until the color of phenolphthalein is just discharged. Boil to coagulate the precipitate. Filter in a weighed Gooch or fritted-glass crucible. Wash. Dry at 105°C for ½ hour and weigh as silver cyanide.

$$\text{g AgCN} \times 10.8 = \text{oz/gal Ag}$$

Sodium Acetate

Pipette a 25 ml sample into a 500 ml distilling flask. Place a few glass beads on the bottom of the flask. Fit the flask with a two-hole rubber stopper, one hole containing a thermometer and the other a dropping funnel. Connect the side arm of the flask to a condenser. Add 175 ml distilled water and 25 ml concentrated sulfuric acid through the dropping funnel. Distill into a 500 ml Erlenmeyer flask containing 50 ml distilled water. Continue the distillation until the temperature reaches about 127°C. Add more water and continue the distillation until 250 to 300 ml water has been distilled over. Add 0.5 ml sulfo-orange indicator and titrate with N/2 sodium hydroxide to a yellow end point.

$$\text{ml NaOH} \times 0.220 = \text{oz/gal } Na_2C_2H_3O_2$$

Sodium Hydroxide

Pipette a 5 ml sample into a 250 ml Erlenmeyer flask. Add 25 ml distilled water and 0.5 ml sulfo-orange indicator. Titrate with N/10 hydrochloric acid to a yellow end point.

$$\text{ml HCl} \times 0.107 = \text{oz/gal NaOH}$$

Cadmium Cyanide Bath for Sodium Hydroxide. Pipette a 5 ml sample into a 250 ml Erlenmeyer flask. Add 50 ml distilled water and 10 ml 10% sodium cyanide. Add 0.5 ml. sulfo-orange indicator. Titrate with standard N/1 hydrochloric acid until the color changes from orange to yellow.

$$\text{ml HCl} \times 0.0107 = \text{oz/gal NaOH}$$

Sodium Cyanide. Sodium hydroxide can be controlled by conductivity in the presence of sodium cyanide, if the sodium cyanide is determined by titration and a correction for the effect of sodium cyanide on the conductivity is determined.

Sodium Thiocyanate

Pipette a 10 ml sample into a 250 ml Erlenmeyer flask. Add 100 ml distilled water. Add sulfuric acid, drop by drop, until the solution just turns blue litmus red. Add 0.5 ml saturated ferric ammonium sulfate solution. Titrate with standard N/10 silver nitrate solution until the red color disappears.

$$\text{ml } AgNO_3 \times 0.109 = NaCNS$$

Sulfate

Chromic Acid Baths. Pipette a 10 ml sample into a 250 ml beaker. Add 75 ml of a reducing mixture consisting of 15 parts by volume isopropanol, 7 parts concentrated hydrochloric acid and 25 parts glacial acetic acid. Dilute with hot water to between 125 and 150 ml and let stand for at least 1 hour at 60 to 70°C. Filter into a 250 ml beaker and wash the green coloration from the paper with hot water. Heat the filtrate to boiling and add, drop by drop, 10 ml 10% barium chloride solution. Allow to settle for 2 to 4 hours at 60 to 70°C. Filter and wash thoroughly. Ignite the paper and precipitate in a weighed porcelain crucible at bright red heat for about 15 minutes. Cool and weigh.

$$\text{g } BaSO \times 5.64 = \text{oz/gal } H_2SO_4$$

Chromic Acid Ratio. The CrO_3:SO_4 ratio of a chromic acid bath can be determined directly from a plating-range test, if plating range standards are available. This method has the advantage that the bath can be adjusted to the widest plating range by test rather than relying on chemical analysis for SO_4 to produce the widest plating range.

Tin

Alkaline Tin Bath. Tin can be determined in an alkaline tin bath by using a rapid plating and deplating method for the measurement of cathode efficiency. If such a method is used, the temperature and caustic concentration effects will have to be taken into account.

Stannous Tin in an Acid Tin Bath. Pipette a 10 ml sample into a 500 ml Erlenmeyer flask. Add 150 ml distilled water, 20 ml concentrated hydrochloric acid, and 2 ml starch solution. Titrate with N/10 iodine solution.

$$\text{ml I} \times 0.0795 = \text{oz/gal Sn}$$

Total Tin in an Acid Bath. Pipette a 10 ml sample into a 500 ml Erlenmeyer flask. Add 150 ml distilled water, 10 ml concentrated hydrochloric acid and 1 g, powdered iron. Attach an outlet tube and boil for 30 minutes to dissolve the iron and reduce the tin. Place the flask in a cooling bath with the outlet tube submerged in saturated sodium bicarbonate solution. When cool, remove the outlet tube, add starch and titrate with N/10 iodine solution.

$$\text{ml I} \times 0.0795 = \text{oz/gal Sn}$$

ZINC

Cyanide Solution No. 1. Pipette a 5 ml sample into a 25 ml volumetric flask. Make up to the mark with methyl alcohol. Shake and allow the crystals to settle. Filter dry into a beaker. Pipette 10 ml filtrate into a 100 ml beaker. Add 10 ml 1-2 hydrochloric acid and 40 ml water. Add 1 ml 10% sodium sulfide. Titrate with N/10 potassium ferrocyanide.

$$\text{ml } K_4Fe(CN)_6 \times 0.438 = \text{oz/gal Zn}$$

Cyanide Solution No. 2. Pipette a 5 ml sample into a 250 ml beaker. Add 25 ml distilled water and 10 ml 15% sodium sulfide solution. Heat to boiling and filter off the precipitate. Wash with hot water and discard the filtrate. Transfer the filter paper and precipitate into the original beaker. Add 10 ml concentrated hydrochloric acid and 1-2 g sodium sulfide crystals. Boil to expel the hydrogen sulfide. Dilute to 150 ml with distilled water and heat to boiling. Titrate with N/10 potassium ferrocyanide using 5% uranium nitrate as an outside indicator. The end point is reached when the addition of 1 drop of the solution being titrated to 2 drops of uranium nitrate on the spot plate produces a reddish-brown precipitate.

$$\text{ml } K_4Fe(CN)_6 \times 0.175 = \text{oz/gal Zn for N/10}$$

REFERENCES

1. George Jernstedt, *Trans. Electrochem. Soc.*, **82,** 135 (1942).
2. A. Dolance and P. W. Healy, *Ind. Eng. Chem. Anal. Ed.*, **17,** 718 (1945).
3. E. A. Parker, *Monthly Rev., Am. Electroplaters' Soc.*, **34,** 33 (1947).
4. H. E. Zentler-Gordon and E. R. Roberts, *Trans. Electrochem. Soc.*, Preprint 90.92 (1946).
5. V. S. Baleya, N. I. Puchenkina and I. A. Korshunov, *Zavodskaya Lab.*, **11,** 644 (1946).
6. Monaweck, *Trans. Electrochem. Soc.*, **82,** 59 (1942).

7. G. Jernstedt, *Trans. Electrochem. Soc.*, **82,** 137 (1942).

BIBLIOGRAPHY

Modern Electroplating, The Electrochemical Society.
Plating and Finishing Guidebook, Metals and Plastics Publications, Inc.
Simple Methods of Analyzing Plating Solutions, Hanson-Van Winkle-Munning Co.
Year Book, Newark Branch, American Electroplaters' Society.
Scott, *Standard Methods of Chemical Analysis*, D. Van Nostrand Co.

APPENDIX

Conversion Factors

From	*To*	*Multiply by*
g/l	oz/gal	0.134
oz/gal	g/l	7.5
amp/sq ft	amp/sq dm	0.108
amp/sq dm	amp/sq ft	9.29
g/l	oz/cu ft	1.00
in.	cm	2.54
cm	in.	0.394
gal	l	3.78
gal	cu ft	0.134
l	gal	0.264
l	cu ft	0.0353
cu ft	l	28.3
cu cm	cu in.	0.0610
cu in.	cu cm	16.4
sq cm	sq in.	0.155
sq in.	sq cm	6.45

Electrochemical Yields at 100% Cathode Efficiency

Metal	*Symbol*	*Atomic weight*	*Density*	*Valence*	*Grams per faraday*	*Ampere-hours per pound*
Aluminum	Al	27.0	2.70	3	9.0	1,352
Cadmium	Cd	112.4	8.65	2	56.2	217
Chromium	Cr	52.0	7.1	6	8.7	1,398
Cobalt	Co	58.9	8.9	2	29.5	412
Copper	Cu	63.6	8.9	1	63.6	191
Copper	Cu	63.6	8.9	2	31.8	383
Gold	Au	197.2	19.3	1	197.2	62
Gold	Au	197.2	19.3	3	65.7	185
Hydrogen	H	1.008	...	1	1.0	12,170
Indium	In	114.8	7.3	3	38.3	318
Iron	Fe	55.9	7.9	2	28.0	435
Lead	Pb	207.2	11.3	2	103.6	118
Magnesium	Mg	24.3	1.74	2	12.2	997
Nickel	Ni	58.7	8.9	2	29.4	414
Oxygen	O	16.0	...	2	8.0	1,521
Platinum	Pt	195.2	21.4	4	48.8	252
Rhodium	Rh	102.9	12.4	3	34.3	355
Silver	Ag	107.9	10.5	1	107.9	113
Tin	Sn	118.7	7.3	2	59.4	205
Tin	Sn	118.7	7.3	4	29.7	410
Tungsten	W	183.9	19.1	6	30.7	396
Zinc	Zn	65.4	7.1	2	32.7	372

Electrochemical Formulas

$$\frac{\text{atomic weight}}{\text{valence}} = \text{equivalent weight}$$

Equivalent weight = grams per Faraday

faraday = 96,500 ampere-seconds = 26.81 ampere-hours

$$\frac{454}{\text{equivalent weight}} \times 26.81 = \text{ampere-hours per pound}$$

$$\frac{\text{equivalent weight}}{26.81} = \text{grams per ampere-hour}$$

density × 2.36 = grams per mil (one square foot)

$$\frac{\text{grams per mil}}{\text{grams per ampere-hour}} = \text{ampere-hours per mil (one square foot)}$$

Electrochemical Equivalents

			One square foot		One square inch
Symbol	*Valence*	*Grams per ampere-hour*	*Ampere-hours per mil*	*Grams per mil*	*Grams per mil*
Au	1	7.35	6.2	45.5	0.316
Au	3	2.45	18.6		
Ag	1	4.03	6.2	24.8	0.172
Cd	2	2.10	9.7	20.6	0.143
Cr	6	0.32	51.8	16.8	0.116
Cu	1	2.37	8.9	21.0	0.146
Cu	2	1.19	17.8		
Fe	2	1.04	17.9	18.7	0.130
In	3	1.43	12.2	17.5	0.121
Ni	2	1.10	19.0	21.0	0.146
Pb	2	3.87	6.9	26.9	0.187
Rh	3	1.28	22.9	29.2	0.203
Sn	2	2.21	7.8	17.2	0.120
Sn	4	1.11	15.6		
Zn	2	1.18	14.3	16.8	0.116

Single Electrode Potentials

Common plating-bath concentrations at room temperature. Hydrogen=0.00

Electrode	*Volts, acid baths*	*Volts, cyanide or alkaline baths*
Ag/Ag^{+}	−0.75	0.39
Cd/Cd^{++}	0.39	0.93
Cn/Cn^{++}	−0.36	0.65
Fe/Fe^{++}	0.20	0.25
In/In^{+++}	0.30	0.65
Ni/Ni^{++}	−0.20	0.34
Sn/Sn^{++}	0.20	
Sn/Sn^{++++}		0.79
Pb/Pb^{++}	0.15	0.55
Zn/Zn^{++}	0.79	1.15

Stripping Chart

The numbers in the chart refer to the solutions in the Stripping Table below.

Metal to be Stripped	*Base Metal*					
	Aluminum alloy	*Brass*	*Copper*	*Nickel*	*Steel*	*Zinc alloy*
Brass					1,26	
Cadmium		2,9,24	2,9,24		1,2,9,24	
Chromium	21	3	3	3	3,5,12	
Copper	1,19	14			1,13,18	15,16,17,18
Gold		25,28	25,28		1,8	28
Lead		27	27		5,27	
Nickel	6	4,6			6,7,11	6
Platinum				22	22	
Rhodium		4,23		4,23	4,23	
Silver	19	20			1	
Tin		2,10	2,10		2,5	
Zinc	19	3	3		1,3,9	

Stripping Table

No.	*Formula*	*Concentration*
1	Sodium cyanide	5 to 10%
	voltage	2 to 6
2	Hydrochloric acid	20 to 30%
	Antimony trioxide	1 to 2%
3	Hydrochloric acid	5 to 15%
	temperature	140 to 175°F
4	Hydrochloric acid	2 to 5%
	voltage	2
5	Sodium hydroxide	5 to 10%
	voltage	2 to 6
6	Sulfuric acid	67%
	voltage	6 to 12
	current density	100 asf

30 to 40 minutes per mil of nickel

No.	*Formula*	*Concentration*
7	Sulfuric acid	65%
	voltage	2
	lead cathodes	
8	Sulfuric acid	98%
	temperature	below 100°F
	anodic treatment	
9	Ammonium nitrate	10 to 15%
10	Ferric chloride	6 to 8%
	Copper sulfate	9 to 10%
	Acetic acid	30 to 50%
	reactivate with hydrogen peroxide	
11	Nitric acid	70%
	Hydrochloric acid	0.1%
12	Sodium carbonate	5%
	temperature	150°F
	current density	20 asf
13	Chromic acid	35%
	Sulfuric acid	5%
14	Sodium sulfide	10 to 20%
	Sulfur	0.7 to 1.5%
	boil to dissolve sulfur	
15	Sodium sulfide	8%
	Sulfur	3%
	temperature	185°F
	boil to dissolve sulfur	
16	Chromic acid	17 to 37%
	temperature	68 to 77°F
	current density	65 to 130 asf
	alternating current 60 cps	
17	Sodium sulfide	10%
	voltage	2
18	Sodium hydroxide	9%
	Sulfur	14%

No.	*Formula*	*Concentration*
19	Nitric acid	50 to 70%
20	Nitric acid	1 vol
	Sulfuric acid	19 vols
	temperature	180°F
	keep work dry entering solution	
21	Same as solution No. 16 but using anodic direct current or an available chromium plating bath.	
22	Hydrochloric acid	2 vols
	Nitric acid	1 vol
	water	1 vol
	This solution will spontaneously decompose so that the life is short. It will also attack the base metal, but it may be used for thin deposits.	
23	Molten sodium cyanide	
	temperature	1100°F
24	Ammonium nitrate	8 to 12%
25	Sodium cyanide	10%
	Hydrogen peroxide (100 vol)	3%
26	Ammonium hydroxide	5 vols
	Hydrogen peroxide (100 vol)	3 vols
27	Acetic acid	5 vols
	Hydrogen peroxide (100 vol)	1 vol
	water	14 vols
28	Conc. Sulfuric acid	100%
	temperature	below 100°F
	make work anodic	

GLOSSARY

The following definitions will be useful in the discussion of plating baths. These definitions are given in conjunction with the topic of this book. Some of them are new and some are simple variations of a subject that requires extensive explanation, e.g. "polarization" and terms related to polaization. The term "free cyanide," for example, is defined only for control purposes and not for the complex reactions that take place in a cyanide plating bath.

Active—Capable of reacting chemically. A metal surface on which an electrodeposit can be bonded.

Adhesion—The tendency of a plate to adhere to the basis metal. Adhesion of a plate may be satisfactory without obtaining bond or perfect adhesion.

Addition Agent—Substances added to a bath to improve the quality of the deposit, extend the plating range or produce a bright plate.

Ampere-Hour—One ampere flowing for one hour. From the number of ampere-hours the quantity of a product formed at an electrode can be calculated.

Anode—The positive electrode.

Anode Sludge.—Compounds formed on the anode due to impurities in the anode or by reaction with substances in the bath.

Atomic—Pertaining to the atom. Atomic or active hydrogen is formed at the cathode, which is unusually reactive before it forms molecular hydrogen.

Atomic Weight—The weight of an element on a scale where oxygen $=16.00$.

Aqueous—A water system.

Balanced Bath—A plating bath in which the amount of metal supplied to the bath is approximately equal to the amount of metal removed.

Base Box—218 square feet.

Base Metal—The major metal in an alloy; such as copper in a copper-base alloy.

Basis Metal—The metal or alloy on which an electroplate is formed.

Bath—The solution used for electroplating.

Bath Voltage—The voltage from the anode to the cathode.

Baumé (Bé)—A specific gravity or density scale. Baumé hydrometers are used to measure the gravity of a solution.

Bent-Cathode Test—A plating-range test using the principle of a bent cathode to obtain a range of current densities.

Bond—A plate held to a basis metal by atomic forces. The bond strength of an electroplate is often greater than the tensile strength of the weaker of the two metals concerned.

Cathode—The negative electrode.

Cathode Efficiency—The percentage of current used to produce the desired product at the cathode.

Chemical Pickling—Removal of the surface layer on a metal by chemical means.

Coating—A finished or protective layer.

Compatible Metals—Metals that are similar. Metals that will codeposit. Metals that will form solid solutions or compounds.

Complex Ion—An ion in which the metal is combined with other elements so that it is not directly charged.

Conductivity—The ability of a bath to conduct current.

Contaminants—Impurities in a plating bath.

Convection Current—A flow of solution set up in a bath during electrolysis due to the change in concentration at the electrodes.

Coulomb—The unit quantity of electricity. One ampere-second.

Crystalline—Property of a surface that shows small crystal faces.

Current—Electric current.

Current Density—The amount of current per unit area. Amperes per square foot.

Cyanide—A chemical radical consisting of carbon and nitrogen. The free cyanide and the combined cyanide together constitute the total cyanide. However, this definition is only for analytical purposes. The combination of cyanide in the plating bath is not always known.

Cyanide (combined).—The amount of cyanide in chemical combination with a platable metal such as copper or zinc to form a complex.

Cyanide (free).—The difference between the combined and total cyanide. The amount of free cyanide as determined by chemical analysis.

Cyanide (metal)—A metallic, salt of hydro cyanic acid, such as copper cyanide or zinc cyanide. Sodium cyanide is an alkali metal cyanide.

Cyanide (total)—The total amount of cyanide (combined plus free) present in a plating bath. The total cyanide as determined by chemical analysis.

Decomposition Potential—The minimum potential at which a reaction at an electrode just starts.

Degreasing—Removal of grease or oil by the use of a solvent.

Deposition Potential—The potential of the cathode against the solution during electrodeposition.

Diffusion Coating—A coating formed by diffusion of one metal into another.

Drag-in—Impurities or salts introduced into a bath with the solution carried by the work into the plating line from the tanks in which it was treated prior to the bath.

Drag-out—Loss of solution from a bath, carried out by the work.

Electrochemical Series—A series of potentials of the metals based on one gram ionic weight. The approximate potentials for simple solutions. Chemical elements arranged in the order of their standard electrode potentials.

Electrodeposition—The deposition of a substance by means of electric current.

Electrode Reaction—A reaction taking place at an electrode during the passage of current.

Electroforming—Electrodeposition over a form to produce a shape that will be removed from the form.

Electrogalvanizing—Production of zinc plate by electrodeposition.

Electrolysis—Chemical decomposition by the action of electric current on an electrolyte.

Electrolyte—A substance which will conduct electricity by means of ions.

Electrolytic—Pertaining to the passage of current across an electrolyte.

Electrolytic Alkaline Cleaning—Cleaning in an alkaline bath by the passage of current. The work to be cleaned may either be cleaned anodically or cathodically.

Electrolytic Cleaning—Cleaning of a metal by electrolysis.

Electrolytic Pickling—Removal of the surface layer of a metal by electrolytic means, the metal being the anode.

Electronegative—A substance that passes to the anode during electrolysis. Anodic.

Electroplating—Plating by electrodeposition.

Electropositive—A substance that passes to the cathode during electrolysis. Cathodic.

Electrorefining—Refining of metals by an electrolytic process.

Electrotinning—Production of tin plate by electroplating.

Equilibrium Potential—The static potential between a metal and a solution of ions of the metal.

Etch—To remove and roughen part of the surface of a metal by the action of a chemical.

Excessive Polarization—Polarization in excess of that normally expected, caused by abnormal local conditions at the electrode.

Faraday—The quantity of current required to deposit 1 gram equivalent weight—96,500 coulombs.

Formulation—The composition and preparation of plating baths.

Gram Atomic Weight—The atomic weight of a substance in grams.

Gram Equivalent Weight—The gram molecular weight divided by the valence.

Gravity—The density of a solution. Usually measured as Baumé gravity.

Hull Test—A plating-range test using the principle of an inclined cathode to obtain a range of current densities.

Hydrogen—A gas given off at the cathode during electrolysis in many electrolytic processes as a result of the electrolytic decomposition of water.

Hydrogen Overvoltage—The voltage required for hydrogen to just begin to develop at the cathode.

Inorganic—Chemical compounds other than organic.

Ion—A charged particle formed by the loss or gain of electrons from a neutral compound in solution.

Ionization—Conversion of a substance into ions.

Limiting Current Density—A maximum current density that cannot be exceeded without sacrificing plate quality. The current density at which a second electrode reaction begins.

Metal Complex—A complex metal salt.

Metal Salt—A salt containing a metal. Metal salts may be formed by the reaction of a metal with an acid.

Mil—One thousandth of an inch (0.001 in.).

Mineral Acids—Commercial inorganic acids, e.g. hydrochloric or sulfuric acid.

Mol (mole)—Gram molecular weight.

Noble Metal—A metal that deposits easily from a plating bath in competition with a second metal. A metal that will deposit exclusively at low current density.

Noble-Metal System—An alloy-plating bath system in which the analysis of the deposit is dependent on the rate of diffusion of the noble metal to the cathode.

Nonaqueous—Other than a water solution, such as a solution in an organic solvent or a fused salt.

Nonconductor—A substance that will not conduct current. A material used to shadow a cathode from the current during plating.

Normal Polarization—An increase in voltage at an electrode due to the passage of current.

Organic—The compounds of carbon.

Overvoltage—The minimum voltage at which a reaction at an electrode just starts to take place. A dynamic potential that is measured and can be distinguished from the decomposition potential as calculated from thermodynamic constants.

Passive—Property of a surface that will not react chemically. Property of a surface on which a bonded electroplate cannot be obtained.

Pickling—Removal of a metal surface layer with acid.

Plate—A layer of metal over a basis metal. An electroplate.

Plating Range—The entire range of current densities over which a satisfactory deposit can be obtained.

Polarization—A reverse potential set up during plating that tends to resist the flow of current. Increase in voltage at an electrode due to the formation of films.

Potential System—An alloy-plating bath system in which the analysis of the deposit is dependent on the ratio of the two soluble metals.

Primary Salt—A salt containing metal or other substances vital to the plating bath. Metal salts, e.g. cyanides are primary salts. Buffers and salts that aid anode corrosion may be considered as secondary salts.

Sandwich Effect—A stressed deposit on both sides of a thin sheet which causes cracking of the sheet.

Solvent—A substance, usually a liquid, capable of dissolving other substances. Organic fluids used to dissolve grease or oil from the surface of metal.

Specification Plating—Plating to meet specifications. Often the customer, e.g., the government, requires that plating meets standards of thickness or corrosion resistance.

Stannic—A tin compound or tin ion of valence 4. The oxidation product of stannous tin.

Stannous—A tin compound or tin ion of valence 2. The reduction product of stannic tin.

Strike—A low cathode efficiency bath used for purposes of obtaining bond of a subsequent plate.

Surface Films—Films formed on the surface of a metal by chemical action.

Trivalent Chromium—Chromium ion formed by reduction of chromic acid during electrolysis.

Valence—The number of charges carried by a simple ion. The number of hydrogen atoms with which one atom of an element will combine.

Vapor Degreasing—Removal of grease or oil from a metal by allowing the vapors of a solvent to condense on the metal.

INDEX

Acid baths, compared to cyanide 118
Acid descaling 63
Acid dip 37
Active surfaces 28
Addition agent 299
Adhesion 299
Analytical methods, for
  acids 278
  ammonia 279
  carbonate 280
  chloride 281
  chromium 281
  copper 281
  cyanide
    free 282
    total 283
  gelatine 284
  general
    cathode efficiency 275
    centrifuge 275
    colorimeter 276
    electrolytic 275
    gravimetric 274
    pH meter 275
    polarograph 276
    volumetric 273
  gold 284
  indium 285
  iron 286
  lead 286
  nickel 287
  Resorcinol 287
  Rochelle salt 287
  silver 288
  sodium acetate 289
  sodium hydroxide 289
  sodium thiocyanate 289
  sulfate 290
  tin 290
  zinc 291
Anode 3
Anodic etching 33
Anodizing of aluminum 78
  Alumilite 79
  architectural 82
  chromic acid 78
  duplex 81
  hard 81
  oxalic acid 80
  sulfuric acid 79
  sulfuric-oxalic 82
Applications of electroplating 242
  appearance 243
  characteristics 246
  compatibility 243
  economic 242
  engineering 243
  experimentation 247
  identification 246
  plating process 243
  salvage of tools 246
  stop-off 249
  table of 244,245
  temporary protection 243,246
  thin deposits 246
  worn parts 246
Automatic plating 258
Alkaline cleaning 48
Alloys, electrochemistry
  controls 234
  diffusion layer 234
  electroplated
    advantages 233
    compatible 229

copper-zinc 232
lead-tin 231
noble metal 231
potential system 231
silver-lead 231
tin-copper 232
limiting current density 234
polarization 234
Alodine 105

Balanced Bath 299
Bath
balanced 3
geometry 28
preparation 25
voltage 300
Blued steel 167
Bonderizing 167
Brass plating 83
bath preparation 85
cathode efficiency 87
characteristics 83
current density 86
off color 87
plating range 86
red 83
white 83
yellow 83
Bright dipping 41
Bronze plating 89
Battelle 90
bright alloy 91
copper-cadmium 89
copper-tin-zinc 91
speculum 91
steel anodes 92

Cadmium plating 95
acid 101
barrel 97
carbonate 100
chemistry 95
control 99
cyanide 96
hydrogen embrittlement 101
limits 96
make-up 99
NaCN/Cd ratio 96
rapid 98
throwing power 98
toxicity 95
Carbide smut 32
Cathode 2
efficiency 9
Chemical displacement 6
Chromate coatings 103
Alodine 105
corrosion resistant 104
Cronak 103
bright dipping 104
paint base 104
Chromium
baths
black 114
bright 108
conventional 107
crack-free 114
hard 108
porous 114
SRHS 113
hard 22
bright 108
plating
anodes 112
control 111
chemistry 106
CrO3/SO4 ratio 112
current density 109
etching 115
operation 108
plating rate 110
plating range 110
thickness 115
troubles 116
Chemical displacement 29
Chemical milling 67
Cleaner
control 57
formulation 52
selection of 56
testing 48
Cleaners
di-phase 48
electrocleaning 54
heavy-duty 49
light-duty 51

medium-duty 50
multipurpose 50
silicate-phosphate 51
soak 54
spray 55
two stage 53
ultrasonic 56
Cleaning, planned 53
Cold casting 20
Compatible metals 300
Conductance, equivalent 225
Conductivity
acids 220
alkalis 220
salts 219
Table 220
Control of plating bath
analytical methods 208
cathode efficiency 212
chemical limits 208
conductivity 211
gravity 211
plating log 213
plating range 209
plating rate 210
rapid control 212
sampling 211
Continuous plating 258
Convection current 300
Convection-stratification 142
Conversion factors 293
Corronizing 236
Coulomb 8
Copper baths, acid
fluoborate 122
pyrophosphate 122
sulfate 118
Copper baths, cyanide
high efficiency 132
plain 129
Rochelle 131
Copper cyanide bath
anodes 134
characteristics 134
contaminants 133
control 133
free cyanide 128
heavy deposits 135
potassium salts 132
strike 130
Copper plating, acid 117
applications 125
electroforming 125
undercoating 125
influence of basis metal 124
preparation for 123
Copper plating, cyanide 127
chemistry 128
preparation for 135
Copper plating for selective carburization 249
acid copper 252
burrs 251
cracks 250
cyanide copper 252
microscopic examination 251
faults 251
poor bond 252
scale 250
Copper sulfate bath 118
anodes 121
control 121
operation 119
preparation 119
Crystal
form 21
structure 20
Crystalline deposits 20
Current
density 3
distribution 10
Cyanide
combined 300
dip 41
free 300
metal 301
total 301

Decomposition potential 301
Degreasing 44
Deposition potential 2,6,301
Deposits, layered 23
Descaling 43
acid 63
mechanical 35

Diffusion
  coatings 237
  alloy bath 234
Double zincate 39

Electrochemical potential series 5
Electrochemical tables
  equivalents 295
  formulas 294
  potentials 295
  Series 301
  yield 294
Electrocleaning 37
Electrode geometry 10
Electrodeposition 18
Electrogalvanizing 198
Electroless plating 165
  nickel 163
    applications 163
    baths 164
    operation 164
    preparation for plating 165
    tin 166
Electrolysis 1
Electrolyte 1
  aqueous 1
Electromotive force series 5
Electronegative 301
Electropolishing 35
Electropositive 302
Electrotinning
  Ferrostan 259
  flow-brightening 261
  lines 260
  thickness 259
Emulsion cleaning 47
Equilibrium potential 302
Equivalent weight 8
Etching cleaner 39
Etch-cleaning 43
Excessive polarization 302

Faraday 8

Glossary 299
Glue 20
Grain size 19

Hardness table 23
Hull test 302
Hydrogen embrittlement 22
  failures 34
Hydrogen overvoltage 6,14,302

Immersion plating 29
Iron plating baths
  chloride 136
  sulfate 138
  sulfate-chloride 138

Layered deposits 23
Layer plating
  alloys 237,240
  chromium-nickel-copper 236
  Corronizing 236
  diffusion alloys 236
    compound formation 240
  continuous solubility 238
    limited solubility 239
    solid solubility 238
  diffusion coatings 237
    rates 239
  duplex nickel 236
  lead-indium 237
  tin-cadmium 236
Lead fluoborate bath
  addition agent 142
  anodes 141
  control 142
  convection 142
  characteristics 141
Lead plating baths
  fluoborate 140
  fluosilicate 140
  sulfamate 143
Lead-tin plating bath
  antimony additions 150
  bearing plate 147
  characteristics 145
  control 148
  copper additions 150
  current density 147
  formulation 146
  gelatine 149
  plating test 150
  solder plate 147
  tin oxidation 148

use 145
Limiting current density 302

Nickel plating 152
control 155,159
duplex 162
preparation for 161
Nickel plating baths
activating 156
bright 155
fine grain 153
fluoborate 158
general purpose 156
hard 153
limits 154
low stress 157
matte 156
striking 156
sulfamate 157
Watts 152
Noble metal 303
impurities 229
plating rate 231
system 231,303

Overvoltage 303
Oxidizing bath, chemical 33
Oxide, natural 64

Painting of
aluminum 271
cadmium 271
steel 270
zinc 271
Painting, preparation for
chemical treatments
acid rinse 268
artificial weathering 268
conversion coatings 267
cleaning 268
etch priming 269
pickling 269
cleaning
barrel 268
flame 268
hand 268
priming
etch 269
metallic pigments 270
red lead 269
wash 269
zinc chromate 269
protection
chemical 266
indoors 266
outdoors 266
severe environments 270
sheltered 266
Parkerizing 167
Parting compound 19
Passivation 30
Phosphate coatings
applications 168
iron phosphate 168
manganese phosphate 168
zinc phosphate 167
Pickles
chrome 62
electrolytic 65
hydrogen peroxide 63
inhibited 64
nitric-hydrofluoric 62
oxidizing 63
Pickling 32
metals 60
terms 65
Pickling acids, mineral
fluoboric 63
hydrochloric 61
nitric 61
phosphoric 62
sulfuric 61
Plastics, plating on
activation 264
ABS 262
copper 264
cleaning 263
electroless copper 264
etching 263
neutralizing 263
sensitizing 264
Plating
quality 14
range 15,303
rate 9
troubles (see *Troubles, plating bath*)

Plating tests
  addition agent 217
  bent cathode 214
  gravity 219
  Hull cell 214
  impurities 217,218
  low current electrolysis 218
  panel reading 217
  primary control 214
  secondary control 214
  slot cell 215
  test panels 217
Polarization 12,303
  excessive 13
Polishing
  chemical 67
  electro- 67
Potential
  decomposition 13
  equilibrium 5
  system 303
Pre-cleaning 45
Preparation of surface for plating
  aluminum 38
  copper alloys 40
  magnesium 42
  steel
    high carbon 33
    low carbon 31
    medium carbon 32
  stainless 35
  zinc alloys 36
Preparation of metals for painting
  (see *Painting*, *preparation for*)
Preparatory steps of plating 25
Pre-plating 43
Primary salt 303
Properties of common plating
  baths 248

Quality plating 25

Rinsing
  automatic 71
  calculations 73
  contamination 70,72
  control 71
  counterflow 74
  efficiency 70,72
  evaluation 75
  experiments 76
  immersion 71
  ratio 73
Robbing 11

Semi-automatic plating 258
Shadowing 11
Silver cyanide baths
  bright 174
  engineering 173
  free cyanide in 170
  high concentration 173
  nitrate-hydroxide 172
  potassium 170
  sodium 170
  strike 175
Silver plating
  applications 178
  control 177
  operation 174
  preparation for 175
Silver properties 169
Solvent cleaning 45
Specific gravity 219,222
  constants 223
  equation 224
  table 220,224
Stainless steel, plating on 158
Stress 21
Strike 304
Striking plating
  double 69
  low efficiency 68
  nickel 69
Striking 29
  zinc alloys 38
Stripping solutions 296
Surface active agents 50

Tin-cadmium alloy 197
Tin characteristics 179
Tin-nickel plating
  bath 192
  operation 193
  properties 192

Tin plating
  acid
    addition agents 182
    anodes 183
    plating rates 180
    plating tests 181
    preparation for 184
    thickness 179
  alkaline
    anodes 189
    bath preparation 186
    cathode efficiency 188
    control 187
    characteristics 186
    plating rate 191
Tin plating baths
  acid
    du Pont Halogen 183
    fluoborate 181
    fluosilicate 183
    sulfate 180
  alkaline
    mixed 189
    potassium 188
    sodium 185
Tin-zinc plating
  alloys 196
  anodes 195
  baths 195,196
  cathode efficiency 195
  corrosion resistance 194
  properties 196
Troubles, plating
  basis metal
    decomposed oil 253
    porous metal 253
  bath
    anodes 256,257
    cleaning 254
    low acid 256
    low cathode efficiency 256
    low cyanide 256
  deposit
    adhesion 255
    blistered 255
    color 254
    distribution 257
    rough 255
  testing
    filtration 254
    hand cleaning 254
    low current density 254

Vapor degreasing 45
Voltage, bath 221,226
  alkaline cleaners 221
  alkaline tin 222
  equation 226
  operation 227
  polarization 227
  Table 227
  Watts nickel 226

Water-break-free surfaces 44

Zincating of aluminum 38
Zinc bath
  acid 198
    control 199
    characteristics 200
    formulation 198
    operation 199
    preparation for plating 200
  cyanide 201
    bath preparation 205
    comparison to acid bath 201
    chemistry 202
    control 205
    formulation 202
    operation 205
    NaCN/Zn ratio 204
    plating limits 206
    preparation for plating 207
Zinc baths
  cyanide
    bright 203
    plain 202
    zinc mercury 203
  acid
    sulfate 198